67

Fortschritte der Chemie organischer Naturstoffe

Progress in the Chemistry of Organic Natural Products

Founded by
L. Zechmeister

Edited by
W. Herz, G. W. Kirby,
R. E. Moore, W. Steglich,
and Ch. Tamm

Authors:
A. A. L. Gunatilaka, Ch. Tamm,
P. Walser-Volken

SpringerWienNewYork

Prof. W. Herz, Department of Chemistry,
The Florida State University, Tallahassee, Florida, U.S.A.

Prof. G. W. Kirby, Chemistry Department,
The University, Glasgow, Scotland

Prof. R. E. Moore, Department of Chemistry,
University of Hawaii at Manoa, Honolulu, Hawaii, U.S.A.

Prof. Dr. W. Steglich, Institut für Organische Chemie der Universität
München, München, Federal Republic of Germany

Prof. Dr. Ch. Tamm, Institut für Organische Chemie der Universität Basel,
Basel, Switzerland

© 1996 by Springer-Verlag/Wien
Softcover reprint of the hardcover 1st edition 1996

Library of Congress Catalog Card Number AC 39-1015

Typesetting: Macmillan India Ltd., Bangalore-25

Printed on acid free and chlorine free bleached paper

With 28 Figures and 1 coloured Plate

ISSN 0071-7886
ISBN-13: 978-3-7091-9408-9   e-ISBN-13: 978-3-7091-9406-5
DOI: 10.1007/978-3-7091-9406-5

Contents

List of Contributors VII

Triterpenoid Quinonemethides and Related Compounds (Celastroloids).
By A. A. L. GUNATILAKA.   1

1. Introduction   2

2. General Structural Features and Nomenclature   4

3. The Families of Celastroloids   6
   3.1. Quinonemethide Triterpenoids   6
   3.2. 14(15)-Enequinonemethide Triterpenoids   7
   3.3. 9(11)-Enequinonemethide Triterpenoids 11
   3.4. Phenolic and 6-Oxophenolic Triterpenoids 14
   3.5. 7-Oxoquinonemethide Triterpenoids 14
   3.6. Dimeric Celastroloids 15
   3.7. Miscellaneous Celastroloids 18

4. Natural Occurrence 18
   4.1. Taxonomic Considerations 18
   4.2. Plant Sources of Celastroloids 19
   4.3. Distribution of Natural Celastroloids 24
   4.4. Celastroloids from Tissue Cultures 24

5. Derivatives of Celastroloids 27

6. The Chemistry of Celastroloids 35
   6.1. Isolation Techniques 35
   6.2. Structure Elucidation 37
       6.2.1. Early Structural Studies of Celastrol and Pristimerin 37
       6.2.2. Application of Spectroscopic Techniques 42
           6.2.2.1. UV/VIS and ORD/CD Spectroscopy 42
           6.2.2.2. Infrared Spectroscopy 47
           6.2.2.3. NMR Spectroscopy 48
               6.2.2.3.1. ^{1}H-NMR Spectroscopy 49
               6.2.2.3.2. ^{13}C-NMR Spectroscopy 59
           6.2.2.4. Mass Spectrometry 70
           6.2.2.5. X-Ray Crystallography 74
   6.3. Chemical Reactions 76
       6.3.1. General Chemical Characterization 76
       6.3.2. Degradation and Oxidation 76
       6.3.3. Reduction and Derivatization 79

    6.3.4. Addition Reactions   82
    6.3.5. Rearrangements   83
    6.3.6. Photochemistry   88

7. Partial Synthesis   89

8. Biosynthetic Aspects   91

9. Biological Activity   95
    9.1. Antimicrobial Activity   96
    9.2. Antitumor Activity   99
    9.3. Other Biological Activities 103

10. Conclusions 103

Addendum 105

Acknowledgements 114

References 114

The Spirostaphylotrichins and Related Microbial Metabolites.
By P. WALSER-VOLKEN and CH. TAMM 125

1. Introduction 125

2. Isolation of Spirostaphylotrichin A (1), B (2), C and D (3/4), Q (5), R (6) and
   F (7) from Cultures of *Staphylotrichum coccosporum* 127

3. Biosynthetic Studies 135
    3.1. Feeding Experiments with ^{14}C-, ^{13}C, and ^{2}H-Labelled Precursors 135
        3.1.1. Experiments with ^{14}C-Labelled Precursors 135
        3.1.2. Experiments with ^{13}C- and ^{2}H-Labelled Precursors 136
    3.2. Investigation of Mutant Strains and Isolation of Further
         Spirostaphylotrichins 142
        3.2.1. Mutagen Treatment and Selection of Mutants 142
        3.2.2. The Mutant Strain *P 84* 143
        3.2.3. The Mutant Strain *P 649* 146

4. Investigation of the Epimerization of Spirostaphylotrichins 150

5. Discussion of the Results Regarding the Biosynthesis of the
   Spirostaphylotrichins. 153

6. Synthetic Approaches Towards the Spirostaphylotrichins 156

7. Related Fungal Metabolites: Triticones, Arthropsolides, and Other Compounds   159
    7.1. Triticones. 159
    7.2. Arthropsolides and Related Compounds 160

References. 164

Author Index 167

Subject Index 171

List of Contributors

GUNATILAKA, Dr. A. A. L., Department of Chemistry, Virginia Polytechnic Institute and State University, 3017 Hahn Hall, Blacksburg, Virginia 24061–0212, USA.

TAMM, Prof. Dr. CH., Institut für Organische Chemie, Universität Basel, St. Johanns-Ring 19, CH-4056 Basel, Switzerland.

WALSER-VOLKEN, Dr. P., Institut für Organische Chemie, Universität Basel, St. Johanns-Ring 19, CH-4056 Basel, Switzerland.

Triterpenoid Quinonemethides and Related Compounds (Celastroloids)

A. A. Leslie Gunatilaka, Department of Chemistry,
Virginia Polytechnic Institute and State University,
Blacksburg, Virginia, USA

Contents

1. Introduction 2

2. General Structural Features and Nomenclature 4

3. The Families of Celastroloids 6
   3.1. Quinonemethide Triterpenoids 6
   3.2. 14(15)-Enequinonemethide Triterpenoids 7
   3.3. 9(11)-Enequinonemethide Triterpenoids 11
   3.4. Phenolic and 6-Oxophenolic Triterpenoids 14
   3.5. 7-Oxoquinonemethide Triterpenoids 14
   3.6. Dimeric Celastroloids 15
   3.7. Miscellaneous Celastroloids 18

4. Natural Occurrence 18
   4.1. Taxonomic Considerations 18
   4.2. Plant Sources of Celastroloids 19
   4.3. Distribution of Natural Celastroloids 24
   4.4. Celastroloids from Tissue Cultures 24

5. Derivatives of Celastroloids 27

6. The Chemistry of Celastroloids 35
   6.1. Isolation Techniques 35
   6.2. Structure Elucidation 37
      6.2.1. Early Structural Studies of Celastrol and Pristimerin 37
      6.2.2. Application of Spectroscopic Techniques 42
         6.2.2.1. UV/VIS and ORD/CD Spectroscopy 42
         6.2.2.2. Infrared Spectroscopy 47
         6.2.2.3. NMR Spectroscopy 48
            6.2.2.3.1. ^{1}H-NMR Spectroscopy 49
            6.2.2.3.2. ^{13}C-NMR Spectroscopy 59

      6.2.2.4. Mass Spectrometry 70
      6.2.2.5. X-Ray Crystallography 74
  6.3. Chemical Reactions 76
    6.3.1. General Chemical Characterization 76
    6.3.2. Degradation and Oxidation 76
    6.3.3. Reduction and Derivatization 79
    6.3.4. Addition Reactions 82
    6.3.5. Rearrangements 83
    6.3.6. Photochemistry 88

7. Partial Synthesis 89

8. Biosynthetic Aspects 91

9. Biological Activity 95
  9.1. Antimicrobial Activity 96
  9.2. Antitumor Activity 99
  9.3. Other Biological Activities 103

10. Conclusions 103

Addendum 105

Acknowledgements 114

References 114

1. Introduction

The triterpenoid quinonemethides constitute a relatively small group of unsaturated and oxygenated D:A-*friedo-nor*-oleananes. In nature these *nor*-triterpenoid pigments are found restricted to the higher plant family, Celastraceae (including Hippocrateaceae; see Sect. 4.1). For this reason, BRÜNING and WAGNER (*10*) coined the general name "celastroloids" for this class of compounds, and in this review, this term is used exclusively for all natural triterpenoid quinonemethides and their structural relatives. Celastroloids usually co-occur with other triterpenoid types like the D:A-*friedo*-oleananes and lupanes. Despite their structural complexity and interesting biological activities, no comprehensive review devoted specifically to celastroloids has appeared to date. However, mention has been made of celastroloids in general reviews and handbooks dealing with other triterpenoid types (*1, 26, 108, 123*), terpenoids and steroids of Sri Lankan plants (*80*), constituents and phytochemistry of Celastraceae (*10, 63, 70, 109*), triterpene phenoldienones (*110*), phenolic triterpenoids of Sri Lankan Celastraceae (*42, 84*), and natural quinonemethides (*16, 143, 144*). Since approximately 70 celastroloids are presently known, a review devoted entirely to these is now appropriate.

This review covers the literature included in Chemical Abstracts up to June 1994 and contains comprehensive discussions on distribution, structure elucidation, chemistry, biological activity and biosynthetic aspects of celastroloids. It also includes all their derivatives and related structures of semisynthetic origin and some unpublished data from our laboratory and those of others. Some conclusions on current status and future prospects are drawn at the end.

The discovery of celastroloids and the history of their structure elucidation demand comment. The discovery of the most common triterpenoid quinonemethides, celastrol (1) and pristimerin (11), and the studies which led to the establishment of their interrelationship will be considered here whereas the history of their structure elucidation will be discussed in Sect. 6.2.1.

The powdered roots of *Tripterygium wilfordii* Hook. f. (family Celastraceae), called "lei king teng" ("thunder god wine") in Chinese, have been used in China for centuries as an insecticide. In 1930 the "thunder god wine" suddenly came into prominence in Chekiang province, China, as the result of a dispute regarding the damage to valuable and highly productive valley lands caused by rain water from the nearby hills following the harvest of *Tripterygium* roots, which left the soil loose, hence flooding the valley. The land owners of the valley had requested that the cultivation of *T. wilfordii* be forbidden. This quickly provoked a strong opposition from vegetable growers who complained that they could not grow their crops without using the powdered roots of *T. wilfordii* to kill noxious insects. Entomologists and other experts were sent to investigate the validity of this claim. As a result of these investigations, CHOU and MEI published the first paper on the root constituents of *Tripterygium wilfordii* in 1936 (*15*). They were able to isolate the insecticidal principle (suspected to be an alkaloid) and dulcitol along with an insecticidally inert pigment named tripterine. Tripterine was later shown to be identical with celastrol obtained by GISVOLD from another Celastraceae species, *Celastrus scandens* (*47*).

The first scientific paper dealing with celastrol was published in 1939 by GISVOLD who wrote, "One time some investigators thought that the chief pigment found in the outer bark of the root of *Celastrus scandens* was β-carotene. β-Carotene is of great importance because of its use as a standard for vitamin A, and therefore confirmation of its existence in a plant from which it might be easily isolated and purified was deemed advisable." However, to his disappointment, the pigment turned out to be different from β-carotene. In GISVOLD's words, "the bark of the root of *Celastrus scandens* contains no β-carotene. A red pigment has been isolated and named celastrol" (*47*). GISVOLD was perhaps unaware that

BHARGAVA had given the same name to a phytosterol isolated from *Celastrus paniculatus* (5).

Tripterygium wilfordii was introduced into the United States from China in 1941 after its claimed insecticidal activity had been demonstrated (*86*). In 1942 SCHECHTER and HALLER (*132*) reisolated from *T. wilfordii* tripterine previously reported by Chinese workers and showed that it was identical with celastrol isolated by GISVOLD. In the early 1950's BHATNAGAR and DIVEKAR (*6*) were prompted to initiate chemical studies on an Indian Celastraceae species, *Pristimeria indica*, as there was a common belief among the people of North Kanara in India that a teaspoonful of the paste (obtained by rubbing the root of this plant on a stone with a little lime juice) administered orally twice a day for 3 days completely cured respiratory diseases. They were successful in isolating the antibacterial principle from the root bark of *P. indica* which separated as bright orange needles and was named pristimerin. The probable relationship between pristimerin and celastrol was noted by KULKARNI and SHAH (*102*) and later by KAMAT, FERNANDES and BHATNAGAR (*98*). However, it was NAKANISHI's group which firmly established that pristimerin is the methylated derivative of celastrol (*118*).

Since the discovery of celastrol (tripterine) and pristimerin and the establishment of their structural interrelationship, a variety of celastroloids have been encountered in plants of the family Celastraceae (including Hippocrateaceae) and these are listed in Tables 1 through 8.

2. General Structural Features and Nomenclature

In general, celastroloids incorporate the 24-*nor*-D : A-*friedo*-oleanane nucleus (Fig. 1) and invariably contain oxygenated functionalities at C-2

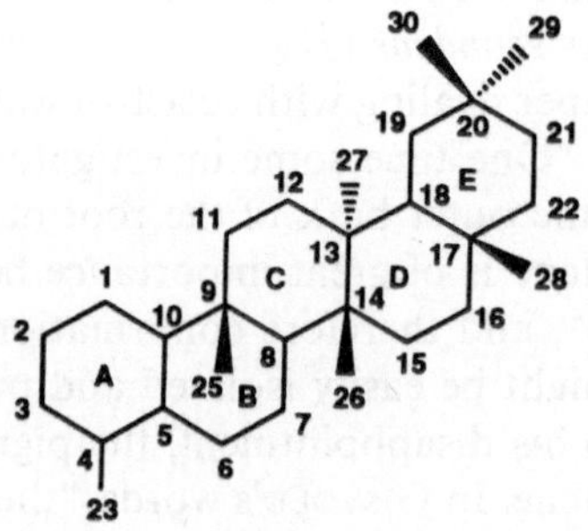

Fig. 1. Basic nucleus of celastroloids

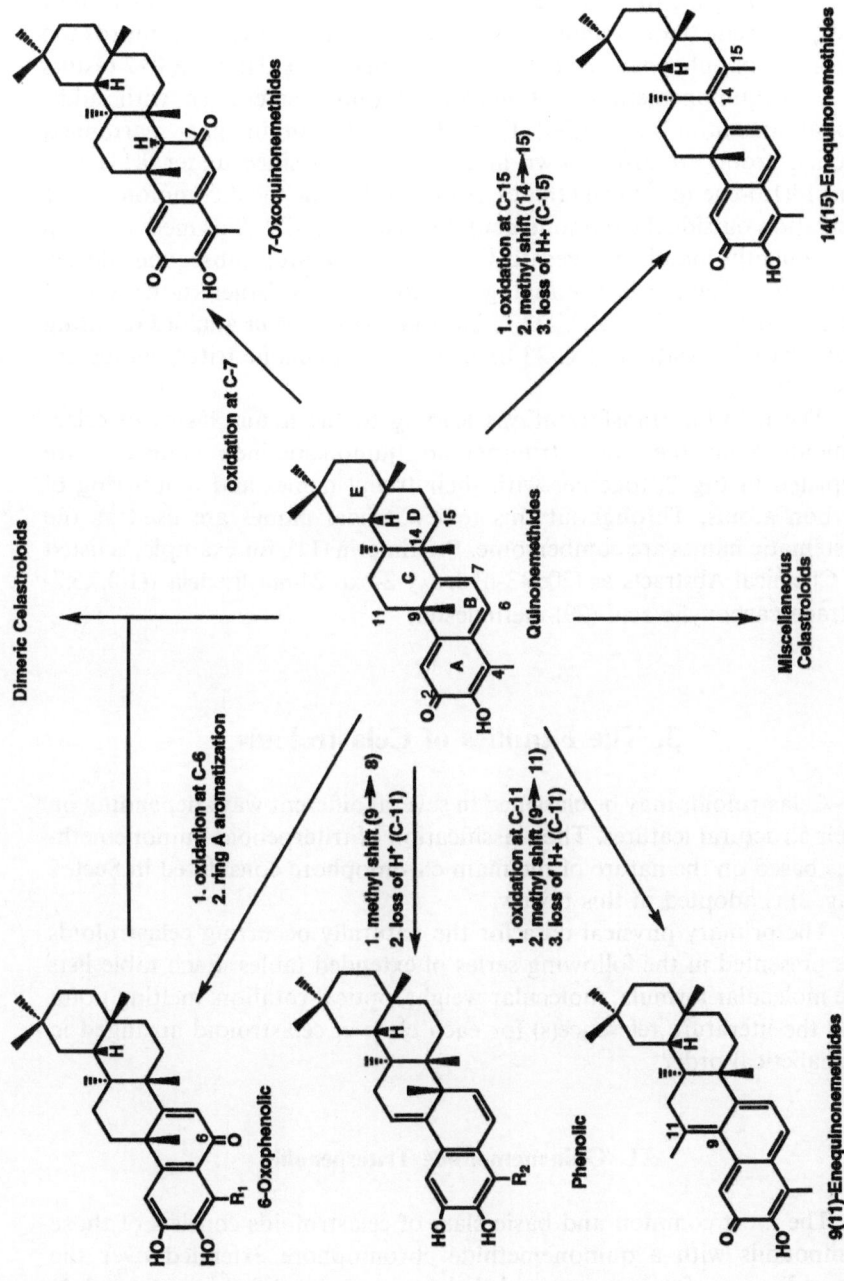

Fig. 2. Major classes of celastroloids and their structural (biogenetic) interrelationship

and C-3, the only exception being celastranhydride (46) in which ring
A has undergone oxidative cleavage. Other known sites of oxidation of
the quinonemethide system are C-6 and C-7. Oxidation at C-6 affords the
class of 6-oxophenolic *nor*-triterpenoids whereas oxidation at C-7 results
in 7-oxoquinonemethides. Triterpenoid quinonemethides with additional unsaturation at C-9(11) or C-14(15) containing a rearranged
methyl group are also known and these are classified under 9(11)-ene-
and 14(15)-ene-quinonemethides, respectively. The most common site of
oxidation outside the quinonemethide system is C-29. This methyl group
is frequently found oxidized to CO_2H (or CO_2Me); subsequent decarboxylation yielding 29-*nor* analogs is also common. Other known sites of
oxidation are C-15, C-21, C-22, C-23, and C-28. 23-*Nor* analogs resulting
from decarboxylation of C-23 oxidized 6-oxophenolic triterpenoids are
also known.

The possible transformations leading to the main classes of celastroloids from the basic triterpenoid quinonemethide congener are
depicted in Fig. 2, together with their trivial names and numbering of
carbon atoms. Throughout this review trivial names are used as the
systematic names are cumbersome. Pristimerin (11), for example, is listed
in Chemical Abstracts as (20α)-3-hydroxy-2-oxo-24-*nor*-friedela-1(10),3,5,7-
tetraen-carboxylic acid-(29)-methylester.

3. The Families of Celastroloids

Celastroloids may be classified in several different ways depending on
their structural features. The classification of triterpenoid quinonemethides based on the nature of the main chromophore considered in Sect. 2
(Fig. 2) is adopted in this review.

The primary physical data for the naturally occurring celastroloids
are presented in the following series of extended tables. Each table lists
the molecular formula, molecular weight, optical rotation, melting point
and the literature reference(s) for each class of celastroloid arranged in
alphabetical order.

3.1. Quinonemethide Triterpenoids

The most common and basic class of celastroloids consists of those
compounds with a quinonemethide chromophore extended over the
A and B rings of the triterpenoid skeleton, as exemplified by celastrol (1)

and pristimerin (**11**). These compounds are listed in Table 1. In all of these, the C-29 methyl group has undergone partial oxidation to $-CH_2OH$ [e.g. excelsine (**3**)] or complete oxidation to $-CO_2H$ [e.g. celastrol (**1**)] sometimes followed by methylation to produce $-CO_2Me$ [e.g. pristimerin (**11**)]. Several examples are also known in which C-29 is lost possibly as a result of a decarboxylation process [e.g. tingenone (**12**)]. Compounds in which the C-29 methyl group has been replaced with a hydroxy group or the C-30 methyl has been oxidized to a hydroxymethyl group are also found in nature [e.g. 20α-hydroxytingenone (**6**) and 30-hydroxypristimerin (**5**), respectively]. These may represent possible biosynthetic precursors to those celastroloids bearing an unsaturation at C-20(21) or C-20(30) positions [see, iguesterin (**13**) and isoiguesterin (**14**) in Table 1; and salaciquinone (**37**) in Table 5].

The two biogenetically related 29-*nor* celastroloids, iguesterin (**13**) and isoiguesterin (**14**) with an endocyclic and exocyclic double bond respectively in ring E are also listed in Table 1. A recent report by CORDELL *et al.* suggested that the structure of 20α-hydroxytingenone (**6**) should be revised to 20-hydroxy-20-*epi*-tingenone (**7**) (*107*). Although the full details have yet to appear, the natural occurrence of the probable biosynthetic precursor of tingenone (**12**) namely, 21-ketopristimerin (**10**) has been reported (*25*). The majority of quinonemethide triterpenoids bear a highly oxidized ring E which contains hydroxy and/or keto functions at C-21 and/or C-22. Only one quinonemethide triterpenoid with a hydroxy group at C-15 is known [15α, 22β-dihydroxytingenone (**2**)] which is suspected to be the biosynthetic precursor of 14(15)-enequinonemethides considered in the next section.

3.2. 14(15)-Enequinonemethide Triterpenoids

The second major group of celastroloids consists of those compounds with a chromophore extended to the ring D of the triterpenoid as a result of an additional double bond located at C-14(15). As a consequence, the methyl group originally present at C-14 has migrated to C-15. Triterpenoids of this class are listed in Table 2.

The majority of compounds belonging to this class are oxidized at C-21, C-22 and C-29 in ring E similar to quinonemethides (see Sect. 3.1). 29-*Nor* analogs are also known. A unique feature of this class of triterpenoids is the presence of a compound oxidized at C-28, namely netzahualcoyol (**20**). It is noteworthy that all 14(15)-enequinonemethides encountered thus far differ only in ring E. It is also interesting that in 21,22-dioxygenated 14(15)-enequinonemethides with an oxo and a hydroxy function [e.g. balaenonol (**16**) and netzahualcoyone (**22**)] the oxo

Table 1. Naturally Occurring Quinonemethide Triterpenoids

Trivial names	R_1	R_2	R_3	R_4	R_5	Molecular formula (MW)	$(\alpha)_D$	mp (°C)	References
Celastrol (1) [= Tripterine]	H	H_2	H_2	CO_2H	Me	$C_{29}H_{38}O_4$ (450.26)	−150° (CHCl$_3$)	204–205	(27, 105, 119, 122)
15α,22β-Dihydroxytingenone (2)	OH	O	β-OH,α-H	H	Me	$C_{28}H_{36}O_5$ (452.26)	−182° (CHCl$_3$)	242–243	(28)
Excelsine (3)	H	β-OH,α-H	H_2	CH_2OH	Me	$C_{29}H_{40}O_4$ (452.26)	−	gum	(11)
21β-Hydroxypristimerin (4)	H	β-OH,α-H	H_2	CO_2Me	Me	$C_{30}H_{40}O_5$ (480.26)	−	238–241	(24)
30-Hydroxypristimerin (5)	H	H_2	H_2	CO_2Me	CH_2OH	$C_{30}H_{40}O_5$ (480.28)	−147°	228–229	(27)
20α-Hydroxytingenone (6)ᵃ	H	O	H_2	OH	Me	$C_{28}H_{36}O_4$ (436.26)	+122° (CHCl$_3$)	207–208.5	(9,37)
20-Hydroxy-20-epi-tingenone (7)	H	O	H_2	Me	OH	$C_{28}H_{36}O_4$ (436.26)	+102° (CHCl$_3$)	202–205	(107)

Compound					Molecular formula (MW)	$[\alpha]_D$	mp (°C)	References
22β-Hydroxytingenone (8) [= Tingenin B]	H	O	β-OH,α-H		Me	$C_{28}H_{36}O_4$ (436.26)	−179° (CHCl₃)	210–211	(120)
Isoiguesterol (9)	H	H₂	CH₂OH		H	$C_{28}H_{38}O_5$ (422.26)	22° (CHCl₃)	118–120	(27)
21-Ketopristimerin (10)	H	O	H₂	CO₂Me	Me	$C_{30}H_{38}O_5$ (478.27)	—	—	(25)
Pristimerin (11)	H	H₂		CO₂Me	Me	$C_{30}H_{40}O_4$ (464.26)	−168° (CHCl₃)	219–220	(6,27,118)
Tingenone (12) [Tingenin A; Maitenin; Maytenin]	H	O	H		Me	$C_{28}H_{36}O_3$ (420.26)	−196° (CHCl₃)	203–204	(120)

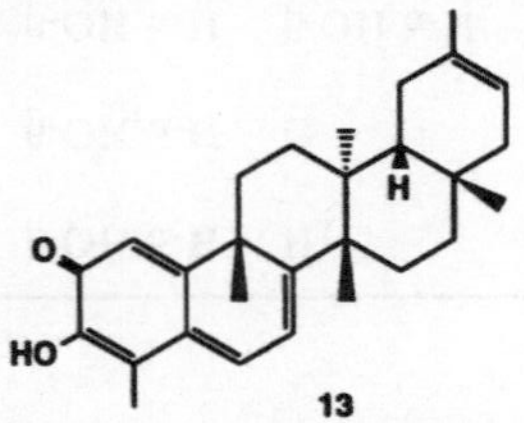

13

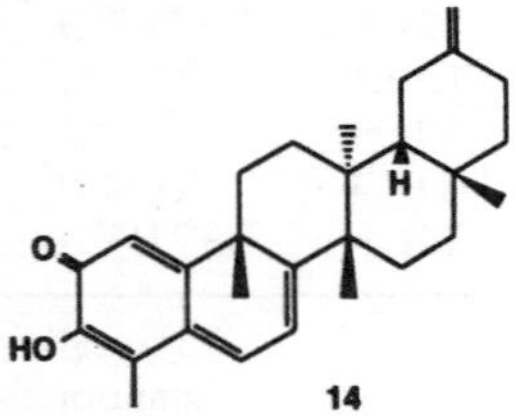

14

Trivial names	Molecular formula (MW)	$(\alpha)_D$	mp (°C)	References
Iguesterin (13)	$C_{28}H_{36}O_2$ (404.27)	−99° (CHCl₃)	196–197	(57)
Isoiguesterin (14)	$C_{28}H_{36}O_2$ (404.27)	−100° (CHCl₃)	203–205	(136)

a Structure of this celastroloid was recently revised to 20-hydroxy-20-*epi*-tingenone (7) (107).

Table 2. *Naturally Occurring 14(15)-Enequinonemethide Triterpenoids*

Trivial names	R_1	R_2	R_3	R_4	R_5	Molecular formula (MW)	$(\alpha)_D$	mp (°C)	References
Balaenol (15)	β-OH,α-H	H_2	Me	H	Me	$C_{28}H_{36}O_3$ (420.27)	+152° (CHCl$_3$)	139–142	(37)
Balaenonol (16)	β-OH,α-H	O	Me	H	Me	$C_{28}H_{34}O_4$ (434.25)	+109° (CHCl$_3$)	205–208	(36, 37)
Isobalaendiol (17)	β-OH,α-H	β-OH,α-H	Me	Me	H	$C_{28}H_{36}O_4$ (436.26)	+65° (CHCl$_3$)	210–212	(37)
Isobalaenol (18)	β-OH,α-H	H_2	Me	Me	H	$C_{28}H_{36}O_3$ (420.27)	–	–	(37)
Netzahualcoyene (19)	H_2	H_2	Me	CO$_2$Me	Me	$C_{30}H_{38}O_4$ (462.28)	–	150–152 and 176–178	(65)
Netzahualcoyol (20)	β-OH,α-H	H_2	CO$_2$Me	CO$_2$Me	Me	$C_{31}H_{38}O_7$ (522.26)	–	amorphous	(65)
Netzahualcoyondiol (21)	β-OH,α-H	β-OH,α-H	Me	CO$_2$Me	Me	$C_{30}H_{38}O_6$ (494.27)	–	amorphous	(65)
Netzahualcoyone (22)	β-OH,α-H	O	Me	CO$_2$Me	Me	$C_{30}H_{36}O_6$ (492.25)	–	210–212	(61, 65)
Netzahualcoyonol (23)	β-OH,α-H	H_2	Me	CO$_2$Me	Me	$C_{30}H_{38}O_5$ (478.27)	–	amorphous	(65)

group is always located at C-22 and the hydroxy at C-21, in contrast to quinonemethide triterpenoids in which the location of these groups is reversed [e.g. 22β-hydroxytingenone (**8**)].

3.3. 9(11)-Enequinonemethide Triterpenoids

In 9(11)-enequinonemethide triterpenoids, the quinonemethide chromophore is found extended to ring C with methyl migration from C-9 to C-11. Three natural triterpenoids with this chromophore are known and these are listed in Table 3. Salacia quinonemethide (**26**) containing a 9(11)-unsaturation is also listed in Table 3, although it bears a contracted ring C.

The structures of 9(11)-enequinonemethide triterpenoids have been questioned on many occasions (*37, 65, 104*) and a critical evaluation of these arguments will be considered in Sect. 10. The exact location of the hydroxy function in hydroxypristimerinene (**24**) has not been determined although it is suspected to be in ring E (*24*).

Table 3. *Naturally Occurring 9(11)-Enequinonemethide Triterpenoids*[a]

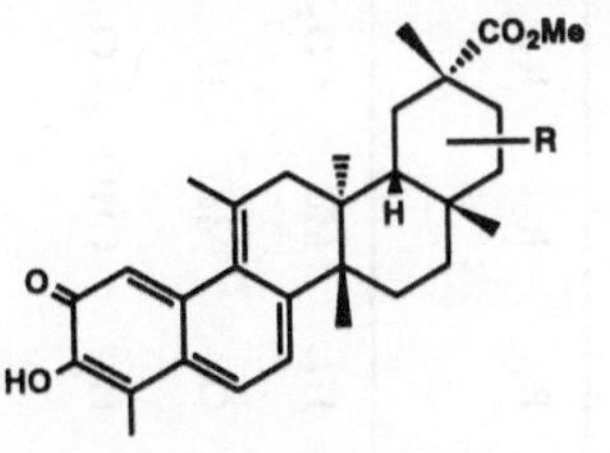

24 R = OH
25 R = H

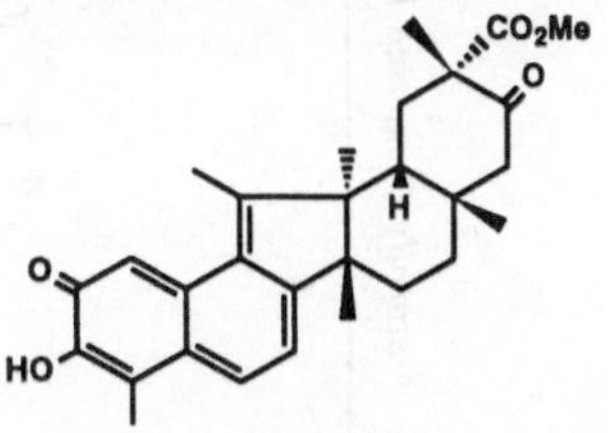

26

Trivial names	R	Molecular formula (MW)	$(\alpha)_D$	mp (°C)	References
Hydroxypristimerinene (**24**)	OH	$C_{30}H_{38}O_5$ (478.27)	–	amorphous	(*24*)
Pristimerinene (**25**)	H	$C_{30}H_{38}O_4$ (462.24)	–	amorphous	(*24*)
Salacia quinonemethide (**26**)		$C_{29}H_{34}O_5$ (462.24)	–	203–205	(*129, 130*)

[a] Structures of these celastroloids have been challenged (see Sect. 10).

Table 4. *Naturally Occurring Phenolic and 6-Oxophenolic Triterpenoids*

Trivial names	R_1	R_2	R_3	Molecular formula (MW)	$(\alpha)_D$	mp (°C)	References
Isopristimerin-III (27)	H_2	Me	CO_2Me	$C_{30}H_{40}O_4$ (464.26)	+ 50° (pyridine)	amorphous	(93)
Isotingenone-III (28)	O	Me	H	$C_{28}H_{36}O_3$ (420.26)	+ 33.1° (pyridine)	amorphous	(93)
23-oxoisopristimerin-III (29)	H_2	CHO	CO_2Me	$C_{30}H_{38}O_5$ (478.27)	–	157–160	(41,79)

Table 4 (continued)

Trivial names	R_1	R_2	Molecular formula (MW)	$(\alpha)_D$	mp (°C)	References
Demethylzeylasteral (30)	CHO	H	$C_{29}H_{36}O_6$ (480.24)	$-67.9°$ (CHCl$_3$)	158–160	(41,43)
Demethylzeylasterone (31)	CO$_2$H	H	$C_{29}H_{36}O_7$ (496.25)	$-36.48°$ (CHCl$_3$)	190–192	(41,47)
23-Nor-6-oxodemethyl pristimerol (32)	H	H	$C_{28}H_{36}O_5$ (452.26)	$-102.0°$ (CHCl$_3$)	228–230	(43)
23-Nor-6-oxopristimerol (33)	H	Me	$C_{29}H_{38}O_5$ (466.27)	$-74.9°$ (CHCl$_3$)	135–138	(43)
Zeylasteral (34)	CHO	Me	$C_{30}H_{38}O_6$ (494.27)	$-136.04°$ (CHCl$_3$)	278–280	(41)
Zeylasterone (35)	CO$_2$H	Me	$C_{30}H_{38}O_7$ (510.26)	$-75.4°$ (CHCl$_3$)	240–242	(78)

3.4. Phenolic and 6-Oxophenolic Triterpenoids

Zeylasterone (**35**) was the first celastroloid in this class to be described (*76*, *78*). It contains a 6-oxo group in addition to the aromatic (phenolic) A ring. Since then a variety of phenolic triterpenoids have been isolated and characterized. Simple phenolic triterpenoids, some of which were previously known as rearrangement products of quinonemethide triterpenoids (see Sect. 6.3.5), have been encountered as natural products and these are listed in Table 4.

All 6-oxophenolic triterpenoids known thus far have been isolated from *Kokoona zeylanica* (*41*, *46*, *77*). However, recently several dimeric celastroloids containing 6-oxophenolic triterpenoid monomers [e.g. cangorosin B (**41**) and umbellatin α (**44**); Table 7] have been described from *Maytenus ilicifolia* (*92*) and *M. umbellata* (*72*). The natural monomeric 6-oxophenolic triterpenoids are listed in Table 4.

3.5. 7-Oxoquinonemethide Triterpenoids

Unlike 6-oxophenolic triterpenoids, their biogenetic relatives the 7-oxoquinonemethide triterpenoids, are rare in nature. Only two belonging to this category, namely, dispermoquinone (**36**) and salaciquinone (**37**), are known (Table 5).

Table 5. *Naturally Occurring 7-Oxoquinonemethide Triterpenoids*

Trivial names	Molecular formula (MW)	$(\alpha)_D$	mp (°C)	References
Dispermoquinone (**36**)	$C_{30}H_{40}O_5$ (480.29)	$-263°$ (CHCl$_3$)	255–257	(*112*)
Salaciquinone (**37**)	$C_{28}H_{36}O_3$ (420.27)	$-130°$ (CHCl$_3$)	252–254	(*142*)

3.6. Dimeric Celastroloids

A number of dimeric celastroloids, all exhibiting some biological activity, have been recently encountered in two plants belonging to *Maytenus* of Celastraceae. Three structural types are known and in all of these the monomer units are linked by an oxygen bridge. In cangorosin A (38), atropocangorosin A (39) and dihydroatropocangorosin A (40) both monomers are of the phenolic triterpenoid type and the linkage is C(3)–O–C(2). Cangorosin B (41) belongs to the dimeric quinonemethide-6-oxophenolic type in which the linkage is the same as in the first type. The third type of dimeric quinonemethide-6-oxophenolic triterpenoids to which the rest of the dimeric triterpenoids belong contains a C(4)–O–C(2) linkage. The dimeric celastroloids are listed in Tables 6 and 7.

Table 6. *Naturally Occurring Dimeric Phenolic Triterpenoids*

Trivial names	Molecular formula (MW)	$(\alpha)_D$	mp (°C)	References
Cangorosin A (38)	$C_{60}H_{84}O_9$ (948.61)	+ 237.4° (CHCl$_3$)	powder	(92)
Atropocangorosin A (atropisomer of 38) (39)	$C_{60}H_{84}O_9$ (948.61)	+ 76.4° (CHCl$_3$)	powder	(92)
Dihydroatropocangorosin A (6',7'-dihydro derivative of 39) (40)	$C_{60}H_{86}O_9$ (950.63)	+ 87.3° (CHCl$_3$)	powder	(92)

Table 7. Naturally Occurring Dimeric Quinonemethide-6-oxophenolic Triterpenoids

Trivial names	Molecular formula (MW)	$(\alpha)_D$	mp (°C)	References
Cangorosin B (41)	$C_{58}H_{74}O_8$ (898.54)	+ 483.3° (CHCl$_3$)	amorphous powder	(92)

Table 7 (*continued*)

Trivial names	R_1	R_2	R_3	Molecular formula (MW)	$(\alpha)_D$	mp (°C)	References
Rzedowskia bistriterpenoid-I[a] (42)	β-a, α-Me	H_2	CO_2Me	$C_{60}H_{78}O_9$ (942.56)	−143.6° (CHCl$_3$)	–	(69)
Rzedowskia bistriterpenoid-II[a] (43)	α-a, β-Me	H_2	CO_2Me	$C_{60}H_{78}O_9$ (942.56)	+260° (CHCl$_3$)	–	(69)
Umbellatin α (44)	β-a, α-Me	O	H	$C_{58}H_{74}O_8$ (898.54)	−187.3°	amorphous	(72)
Umbellatin β (45)	α-a, β-Me	O	H	$C_{58}H_{74}O_8$ (898.54)	+196.6°	amorphous	(72)

[a] Compound not named in the original paper.

3.7. Miscellaneous Celastroloids

Celastranhydride (**46**) which belongs to none of the previous groups, has been isolated from *Kokoona zeylanica* and detected in several other species of Celastraceae (*46*) is listed in Table 8. Although celastranhydride lacks the typical quinonemethide chromophore, it is suspected to

Table 8. *Miscellaneous Naturally Occurring Celastroloids*

Trivial names	Molecular formula (MW)	$(\alpha)_D$	mp (°C)	References
Celastranhydride (**46**)	$C_{28}H_{36}O_5$ (452.26)	+ 160.8° (CHCl₃)	amorphous	(*41, 46*)

be formed from a quinonemethide congener as the result of ring A oxidation followed by cyclization (see Sect. 8). Related anhydrides [e.g. **77**] formed as a result of chemical oxidation of both A and B rings are known (see Scheme 5, Sect. 6.3.2).

4. Natural Occurrence

4.1. Taxonomic Considerations

As noted in the introduction celastroloids have been encountered exclusively in plants of the families Celastraceae and Hippocrateaceae of the angiosperm order Celastrales. The family Celastraceae, commonly known as the bittersweet family, consists of 50–55 genera and 800–850 species distributed pantropically, but with a fair number of species in temperate regions. *Maytenus* (225 species), *Euonymus* (200 species), *Cassine* (40 species), and *Celastrus* (30 species) are the largest genera (*90, 148*). The family Hippocrateaceae consists of only two genera, *Salacia* (200 species) and *Hippocratea* (100 species) with species widespread in tropical regions (*148*). The taxonomic position of the family Hippocrateaceae is presently being debated. Some taxonomists would combine Hippocrateaceae with the Celastraceae whilst others maintain that it should be considered a family distinct from Celastraceae. ROBSON argues that the Hippocrateaceae are not a natural group, but derived in two separate lines from the Celastraceae (*131*). HEYWOOD has recently submerged all species of Hippocrateaceae into the family Celastraceae (*90*). According to HEGNAUER the amalgamation of Hippocrateaceae into Celastraceae is justified by the presence of some common chemotaxonomic markers such as dulcitol, polyisoprene (gutta-percha) and triterpenoid quinonemethides (*89*). A recent review by RAVELO *et al.* has pointed out some chemotaxonomic relationships between the families Celastraceae and Lamiaceae (*128*).

The Celastraceae (including Hippocrateaceae) are a family of trees and shrubs, many of which are climbing (e.g. *Salacia*) or twining (e.g. *Hippocratea*) in habit. The classification at the generic level has undergone changes due to hybridization (*10*), e.g. genus *Maytenus* now includes the species formerly placed in *Gymnosporia*. Similarly, some *Elaeodendron* species are now included in the genus *Reissantia*. This has resulted in many species having several synonyms, some of which are indicated in Table 10.

4.2. Plant Sources of Celastroloids

The 1978 review of Celastraceae constituents by BRÜNING and WAGNER listed 9 celastroloids including saptarangi quinone-A (**47**) (*10*). Since saptarangi quinone-A is reported to be a naphthoquinone (**47**) (*101*) it is not included in this review. However, since it contains 30 carbon atoms and occurs in *Salacia macrosperma*, a plant species known to contain quinonemethides and other triterpenoids, a reinvestigation of its structure by modern NMR techniques may prove otherwise.

47

Plant source(s) together with the plant part(s) in which celastroloids were encountered are listed in Table 9. The physical characteristics of a red crystalline antibacterial and antitumor substance isolated from *Celastrus articulatus* (*146*) suggested it to be a celastroloid belonging to quinonemethide or enequinonemethide class. Since this substance is not fully characterized it is not included in Table 9. Fifty-four species are known to contain celastroloids whose structures are given in Tables 1 to 8. With each celastroloid the structural type is indicated for easy reference to the Tables. The codes for the various structural classes are as follows; A = quinonemethide (Table 1), B = 14(15)-enequinonemethide (Table 2), C = 9(11)-enequinonemethide (Table 3), D = phenolic (Table 4), E = 6-oxophenolic (Table 4), F = 7-oxoquinonemethide (Table 5), G = dimeric phenolic (Table 6), H = dimeric quinonemethide-6-oxophenolic (Tables 7), I = miscellaneous (Table 8). The codes for plant parts are as follows; ap = aerial parts, bk = bark, bw = bark wood, ks = 'kokum soap' (see Sect. 9), ob = outer bark, orb = outer root bark,

Table 9. *Plant Sources of Celastroloids*

Triterpenoid	Class	Plant source	Plant part	References
Atropocangorosin A (**39**)	G	*Maytenus ilicifolia*	–	(*92*)
Balaenol (**15**)	B	*Cassine balae*	rb	(*36, 37*)
Balaenonol (**16**)	B	*Cassine balae*	rb	(*36, 37*)
Cangorosin A (**38**)	G	*Maytenus ilicifolia*	–	(*92*)
Cangorosin B (**41**)	H	*Maytenus ilicifolia*	–	(*92*)
Celastranhydride (**46**)	I	*Cassine balae*	rb	(*44, 46*)
		Kokoona reflexa	rb	(*46*)
		K. zeylanica	rb	(*44, 46*)
		Reissantia indica	rb	(*44, 46*)
Celastrol (**1**)	A	*Catha cassinoides*	rb	(*57*)
		Celastrus paniculatus	orb	(*46*)
		C. scandens	–	(*47*)
		C. strigillosus	rb	(*119*)
		Kokoona ochrasia	sb	(*122*)
		K. zeylanica	osb	(*41*)
		Maytenus umbellata	rb	(*62*)
		Mortonia greggii	rt	(*29*)
		M. palmeri	rt	(*29*)
		Orthosphenia mexicana	rb	(*65*)
		Salacia reticulata var. β-diandra	rb	(*27*)
		Tripterygium regelii	rb	(*119*)
		T. wilfordii	–	(*15, 132*)

Table 9 (*continued*)

Triterpenoid	Class	Plant source	Plant part	References
Demethylzeylasteral (**30**)	D	*Kokoona zeylanica*	orb	(*41*)
Demethylzeylasterone (= Desmethylzeylasterone) (**31**)	D	*Kokoona zeylanica*	osb	(*41, 77*)
Dihydroatropocangorosin A (**40**)	G	*Maytenus ilicifolia*	–	(*92*)
15α,22β-Dihydroxytingenone (**2**)	A	*Cassine balae*	orb	(*28*)
Dispermoquinone (**36**)	F	*Austroplenckia populnea*	rb	(*137*)
			bw	(*138*)
		Maytenus dispermus	orb	(*112*)
Excelsine (**3**)	A	*Hippocratea excelsa*	sb + rb	(*11*)
Hydroxypristimerinene (**24**)	C	*Prionostemma aspera*	rb	(*24*)
		Salacia sp.	rb	(*24*)
21β-Hydroxypristimerin (**4**)	A	*Salacia sp.*	rb	(*24*)
30-Hydroxypristimerin (**5**)	A	*Salacia reticulata* var. β-diandra	rb	(*27*)
20α-Hydroxytingenone (= 20α-Hydroxymaitenin) (**6**)[a]	A	*Austroplenckia populnea*	rb	(*137*)
		Cassine balae	orb	(*37*)
		Euonymus tingens	sb	(*9*)
		Maytenus horrida	rb	(*65*)
		M. nemerosa	wd + st	(*35*)
		M. rigida	rt	(*111*)
		Rzedowskia tolantonguensis	ap	(*64*)
		Salacia macrosperma	rb	(*130*)
20-Hydroxy-20-*epi*-tingenone (**7**)	A	*Kokoona ochracea*	sb	(*122*)
		Glyptopetalum sclerocarpum	sb	(*107*)
22β-Hydroxytingenone (= Tingenin B) (**8**)	A	*Acanthothamnus aphyllus*	rt	(*34*)
		Cassine balae	rb	(*37*)
		Euonymus tingens	sb	(*120*)
		Glyptopetalum sclerocarpum	sb	(*3*)
		Maytenus laevis	rb	(*60*)
		M. obtusifolia	rt	(*25*)
		Maytenus sp.	–	(*22*)
		Salacia reticulata var. β-diandra	rb	(*27*)
Iguesterin (**13**)	A	*Catha cassinoides*	rb	(*57*)
		C. edulis	rt	(*62*)
		Gymnosporia emarginata	rb, rt	(*147*)
		G. montana	sb, rb	(*97*)
		Maytenus umbellata	rt	(*62*)
		Salacia reticulata var. diandra	osb	(*103*)
Isobalaendiol (**17**)	B	*Cassine balae*	orb	(*37*)
Isobalaenol (**18**)	B	*Cassine balae*	orb	(*37*)
Isoiguesterin (**14**)	A	*Salacia madagascariensis*	rt	(*136*)
		S. reticulata var. β-diandra	rb	(*103, 142*)

Table 9 (continued)

Triterpenoid	Class	Plant source	Plant part	References
Isoiguesterol (9)	A	Salacia reticulata		
		var. β-diandra	rb	(27)
Isopristimerin-III (27)	D	Maytenus ilicifolia	rb	(93)
Isotingenone-III (28)	D	Maytenus ilicifolia	rb	(93)
21-Ketopristimerin (10)	A	Salacia sp.		(25)
Netzahualcoyene (19)	B	Maytenus horrida	rb	(65)
		Salacia reticulata		
		var. β-diandra	rb	(27)
Netzahualcoyol (20)	B	Orthosphenia mexicana	orb	(65)
Netzahualcoyondiol (21)	B	Orthosphenia mexicana	orb	(65)
		Rzedowskia tolantonguensis	ap	(64)
Netzahualcoyone (22)	B	Orthosphenia mexicana	orb	(61,65)
		Rzedowskia tolantonguensis	ap	(64,69)
Netzahualcoyonol (23)	B	Orthosphenia mexicana	orb	(65)
23-Nor-6-Oxodemethylpristimerol (32)	E	Kokoona zeylanica	orb	(43)
23-Nor-6-Oxopristimerol (33)	E	Kokoona zeylanica	orb	(43)
23-Oxoisopristimerin-III (29)	D	Kokoona zeylanica	rb	(41,79)
Pristimerin (11)	A	Acanthothamnus aphyllus	rt	(34)
		Austroplenckia populnea	rb	(138)
		Cassine balae	rb	(37)
		C. metabelica	–	(10)
		Catha cassinoides	rb	(57)
		C. edulis	rb	(4)
		Celastrus dispermus	osb	(73)
		C. paniculatus	orb	(94)
			rb,st	(125)
		Crossopetalum uragoga	rb, rm	(32)
		Denhamia pittosporoides	orb	(73)
		Gymnosporia emarginata	rb, rt	(147)
		G. montana	sb, rb	(97)
		Hippocratea excelsa	sb + rb	(11)
		Kokoona reflexa	orb	(46)
		K. zeylanica	osb, ks	(78)
		Maytenus boaria	rt	(25)
		M. chuchuhuasca	sb	(113)
		M. dispermus	–	(10)
		M. horrida	rb	(65)
		M. ilicifolia	rb	(93)
			rc	(54)
		M. obtusifolia	rt	(25)
		M. rigida	rb	(23)
		M. umbellata	rb	(62)
		Pachystima canbyi	rb	(33)
		Plenckia populnea	rt	(55)

Table 9 (*continued*)

Triterpenoid	Class	Plant source	Plant part	References
		Prionostemma aspera	wd	*(23, 51)*
			rb	*(24)*
		Pristimeria indica	rb	*(6, 7, 8)*
		P. grahamii	–	*(102)*
		Reissantia indica	rb	*(45)*
		Rzedowskia tolantonguensis	rt	*(69)*
		Salacia crassifolia	–	*(56)*
		S. macrosperma	rb	*(130)*
		S. reticulata var. β-diandra	rb	*(27)*
			osb	*(103)*
		Schaefferia cuneifolia	rt	*(30)*
		Zinowiewia integerrima	rb	*(31)*
Pristimerinene (**25**)	C	*Prionostemma aspera*	wd	*(24)*
		Salacia sp.	rb	*(24)*
Rzedowskia bistriterpenoid I (**42**)	H	*Rzedowskia tolantonguensis*	rt	*(69)*
Rzedowskia bistriterpenoid II (**43**)	H	*Rzedowskia tolantonguensis*	rt	*(69)*
Salacia quinonemethide (**26**)	C	*Salacia macrosperma*	rb	*(129, 130)*
Salaciquinone (**37**)	F	*Salacia reticulata* var. β-diandra	rb	*(142)*
Tingenone	A	*Acanthothamnus aphyllus*	rt	*(34)*
(= Tingenin A = Maitenin	A	*Cassine balae*	rb	*(37)*
= Maytenin) (**12**)		*Catha cassinoides*	rb	*(57)*
		C. edulis	rb	*(4)*
		Crossopetalum uragoga	rb, rm	*(32)*
		Euonymus tingens	sb	*(9, 99)*
		Gymnosporia emarginata	rb	*(147)*
		G. montana	sb, rb	*(96, 97)*
		Hippocratea excelsa	sb + rb	*(11)*
		Kokoona ochracea	sb	*(122)*
		Maytenus chuchuhuasca	tb	*(113)*
		M. horrida	rb	*(65)*
		M. ilicifolia	rc	*(54)*
		M. laevis	rb	*(60)*
		M. nemerosa	st + wd	*(35)*
		M. obtusifolia	rt	*(25)*
		M. rigida	rt	*(111)*
		M. umbellata	rt	*(62)*
		Maytenus sp.	rc	*(18, 52)*
		Peritassa campestris	rb	*(19)*
		Plenckia populnea	rb	*(19)*
			rt	*(55)*
		Prionostemma aspera	wd	*(24)*
		Reissantia indica	rb	*(45)*
		Rzedowskia tolantonguensis	ap	*(64)*
		Salacia crassifolia	rt	*(56)*

Table 9 (*continued*)

Triterpenoid	Class	Plant source	Plant part	References
		S. macrosperma	rb	(*130*)
		Schaefferia cuneifolia	rt	(*30*)
Umbellatin α (**44**)	H	*Maytenus umbellata*	rt	(*72*)
Umbellatin β (**45**)	H	*Maytenus umbellata*	rt	(*72*)
Zeylasteral (**34**)	E	*Celastrus paniculatus*	orb	(*46*)
		Kokoona reflexa	orb	(*46*)
		K. zeylanica	osb	(*41, 77*)
			ks	(*41*)
Zeylasterone (**35**)	E	*Celastrus paniculatus*	orb	(*46*)
		Kokoona reflexa	orb	(*46*)
		K. zeylanica	osb	(*41, 77*)
			ks	(*41*)

ª Structure recently revised to 20-hydroxy-20-*epi*-tingenone (**7**) (*107*).

osb = outer stem bark, rb = root bark, rc = root cortex, rm = root
medulla, rt = root, sb = stem bark, st = stem, tb = trunk bark,
wd = wood.

4.3. Distribution of Natural Celastroloids

The distribution of natural celastroloids in various genera and species
of the Celastraceae is presented in Table 10. Out of the 54 species
investigated 35 species contained pristimerin (**11**), 25 species contained
tingenone (**12**) and 14 species contained celastrol (**1**). Nearly all species
reported to contain celastroloids furnished at least one of these three
compounds; thus their presence may be of some chemotaxonomic importance since celastroloids have not been encountered in any plant
species outside Celastraceae (including Hippocrateaceae).

4.4. Celastroloids from Tissue Cultures

As an alternative source of supply of the antileukaemic macrocyclic
alkaloid, maytansine, KUTNEY and his coworkers have investigated its
potential production in tissue cultures of *Maytenus buchananii* (*104*).
Instead of producing maytansine the cultures were found to produce,
among other triterpenoids, the cytotoxic triterpenoid quinonemethides

Table 10. *Distribution of Natural Celastroloids*

Plant Source		Celastroloids Encountered	Reference(s)
Genus	Species[a]		
Acanthothamnus	A. aphyllus	(8), (11), (12)	(34)
Austroplenckia	A. populnea	(6), (11), (36)	(138)
Cassine	C. balae	(2)	(28)
	(= Elaeodendron balae)	(6), (8), (11), (12), (17), (18)	(37)
	C. metabelica		
	(= Catha cassinoides	(15), (16)	(36,37)
	= Maytenus		
	canariensis)	(11)	(57)
Catha	C. cassinoides	(1), (11), (12), (13)	(57)
	(= Cassine metabelica		
	= Maytenus		
	canariensis)		
	C. edulis	(11), (12)	(4)
		(13)	(62)
Celastrus	C. dispermus	(11)	(73)
	(= Maytenus dispermus)		
	C. paniculatus	(1), (34), (35)	(46)
		(11)	(46,94,125)
	C. scandens	(1)	(47,132)
	C. strigillosus	(1)	(119)
Crossopetalum	C. uragoga	(11), (12)	(32)
Denhamia	D. pittosporoides	(11)	(73)
Elaeodendron	E. balae	see Cassine balae	
Euonymus	E. tingens	(6)	(9)
		(8)	(120)
		(12)	(99)
Glyptopetalum	G. sclerocarpum	(8)	(3)
Gymnosporia	G. emarginata	(11), (13)	(147)
	G. montana	(11), (13)	(97)
	(= Maytenus	(12)	(96,97)
	senegalensis)		
Hippocratea	H. aspera	see Prionostemma aspera	
	H. excelsa	(3), (11), (12)	(11)
Kokoona	K. ochrasia	(1), (7), (12)	(122)
	K. reflexa	(34), (35), (46)	(46)
	K. zeylanica	(1), (30)	(41)
		(11)	(78)
		(29)	(41,79)
		(31), (34), (35)	(41,77)
		(32), (33)	(43)
Maytenus	M. boaria	(11)	(25)
	M. canariensis	see Catha cassinoides	
	M. chuchuhuasca	(11), (12)	(113)
	(= M. krukovii)		

Table 10 (*continued*)

Plant Source		Celastroloids Encountered	Reference(s)
Genus	Species[a]		
	M. dispermus	(11)	(10)
	(= Celastrus dispermus)	(35)	(112)
	M. horrida	(8), (11), (12), (19)	(65)
	M. ilicifolia	(11)	(54, 93)
		(12)	(54)
		(27), (28)	(93)
		(38), (39), (40), (41)	(92)
	M. laevis	(8), (12)	(60)
	M. nemerosa	(6), (12)	(35)
	M. obtusifolia	(8), (11), (12)	(25)
	M. rigida	(6), (12)	(111)
	M. umbellata	(1), (12), (13)	(62)
		(44), (45)	(72)
Mortonia	M. greggii	(1)	(29)
	M. palmeri	(1)	(29)
Orthosphenia	O. mexicana	(1), (20), (21), (23)	(65)
		(22)	(61, 65)
Pachystima	P. canbyi	(11)	(33)
Peritassa	P. campestris	(12)	(33)
	(= Salacia campestris)		
Plenckia	P. populnea	(11)	(33)
		(12)	(19, 55)
Prionostemma	P. aspera	(11)	(23, 24, 51)
	(= Hippocratea aspera)	(12), (24), (25)	(24)
Pristimeria	P. grahamii	(11)	(102)
	(= Reissantia grahamii)		
	P. indica	(11)	(6, 8)
	(= Reissantia indica = Hippocratea indica)		
Reissantia	R. indica	(11), (12)	(45)
	(= Hippocratea indica = Pristimeria indica)	(46)	(44, 46)
Rzedowskia	R. tolantonguensis	(6), (12)	(64)
		(11)	(64, 69)
		(42), (44)	(69)
Salacia	S. crassifolia	(11), (12)	(56)
	S. macrosperma	(6), (11), (12)	(130)
		(26)	(129, 130)
	S. madagascariensis	(14)	(136)
	S. reticulata var. diandra	(11), (13)	(103)
	S. reticulata var. β-diandra	(1), (5), (8), (9), (11), (12), (19)	(27)
		(14), (37)	(142)
	Salacia species (unidentified)	(4), (24)	(24)

Table 10 (*continued*)

Plant Source		Celastroloids Encountered	Reference(s)
Genus	Species[a]		
	Salacia species (unidentified)	(10)	(*25*)
Schaefferia	S. *cuneifolia*	(11), (12)	(*30*)
Tripterygium	T. *regelii*	(1)	(*119*)
	T. *wilfordii*	(1)	(*15, 132*)
	T. *wilfordii* (cell culture-tissue line TRP 4a)	(2), (6), (48), (49)	(*105*)
Zinowiewia	Z. *integerrima*	(11)	(*31*)

[a] Synonyms and other details in parenthesis.

tingenone (12) and 22β-hydroxytingenone (8) in equal amounts giving a combined yield of 0.01% of the dry weight of cells.

Several triterpenoid quinonemethides were also obtained as by-products in a large scale shake-flask cultivation of *Tripterygium wilfordii* cell tissue line TRP 4a utilized for the production of the cytotoxic diterpene triptolide (*105*). Triterpenoid quinonemethides identified in the cell culture extract included celastrol (1), tingenone (12) and 20α-hydroxytingenone (6). Two other quinonemethides were also isolated and these were tentatively identified by IR, UV, MS and ^{1}H-NMR spectral data as analogs of celastrol and tingenone with undefined stereochemistry as represented in 48 and 49, respectively.

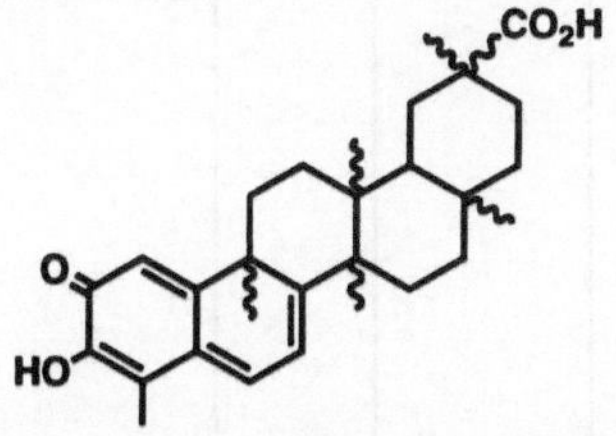

48

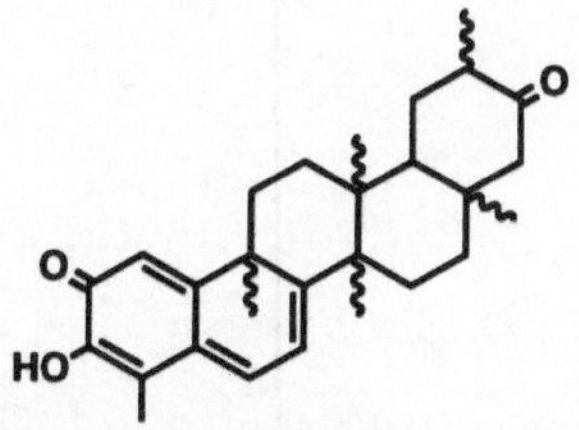

49

5. Derivatives of Celastroloids

Structural investigation of celastroloids, in addition to application of spectroscopic techniques, has relied heavily on the preparation and characterization of derivatives. Some of these derivatives subsequently

Table 11. *Some Derivatives of Celastroloids*

Trivial name(s)	Derived from	Molecular formula (MW)	$(\alpha)_D$	mp (°C)	Spectral data reported	References
21-Deoxyexcelsine (50)	Pristimerin (11) Dispermoquinone (36)	$C_{29}H_{40}O_3$ (436.30)	–	208–210	UV, IR ^{1}H-NMR	(95, 112)

Table 11 (continued)

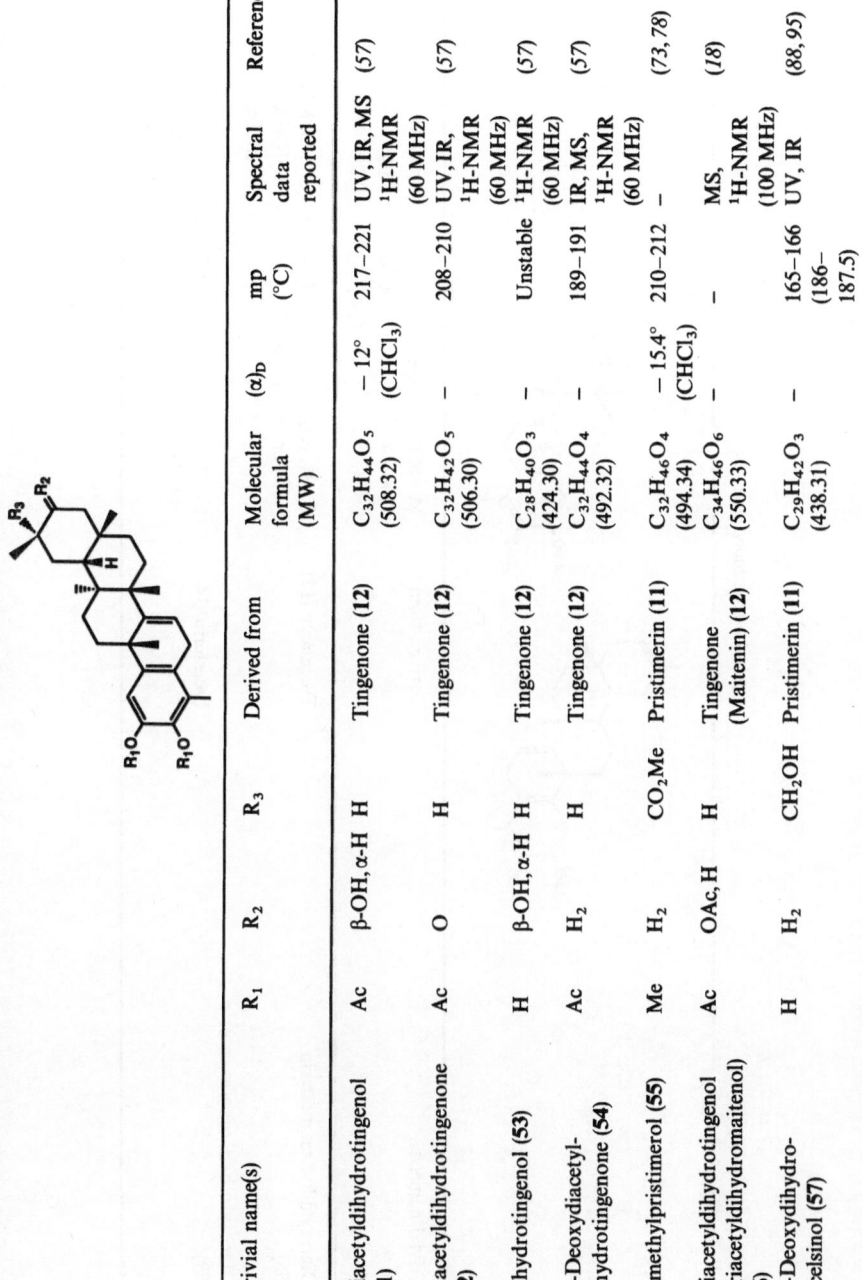

Trivial name(s)	R₁	R₂	R₃	Derived from	Molecular formula (MW)	$(\alpha)_D$	mp (°C)	Spectral data reported	References
Diacetyldihydrotingenol (51)	Ac	β-OH,α-H	H	Tingenone (12)	$C_{32}H_{44}O_5$ (508.32)	−12° (CHCl₃)	217–221	UV,IR,MS ¹H-NMR (60 MHz)	(57)
Diacetyldihydrotingenone (52)	Ac	O	H	Tingenone (12)	$C_{32}H_{42}O_5$ (506.30)	–	208–210	UV,IR, ¹H-NMR (60 MHz)	(57)
Dihydrotingenol (53)	H	β-OH,α-H	H	Tingenone (12)	$C_{28}H_{40}O_3$ (424.30)	–	Unstable	¹H-NMR (60 MHz)	(57)
21-Deoxydiacetyl-dihydrotingenone (54)	Ac	H₂	H	Tingenone (12)	$C_{32}H_{44}O_4$ (492.32)	–	189–191	IR, MS, ¹H-NMR (60 MHz)	(57)
Dimethylpristimerol (55)	Me	H₂	CO₂Me	Pristimerin (11)	$C_{32}H_{46}O_4$ (494.34)	−15.4° (CHCl₃)	210–212	–	(73,78)
Triacetyldihydrotingenol (Triacetyldihydromaitenol) (56)	Ac	OAc, H	H	Tingenone (Maitenin) (12)	$C_{34}H_{46}O_6$ (550.33)	–	–	MS, ¹H-NMR (100 MHz)	(18)
21-Deoxydihydro-excelsinol (57)	H	H₂	CH₂OH	Pristimerin (11)	$C_{29}H_{42}O_3$ (438.31)	–	165–166 (186–187.5)	UV, IR	(88,95)

Table 11 (*continued*)

Trivial name(s)	Derived from	Molecular formula (MW)	$(\alpha)_D$	mp (°C)	Spectral data reported	References
Diacetyldihydroiguesterin (58)	Tingenone (12)	$C_{32}H_{42}O_4$ (490.31)	$-27°$ (CHCl$_3$)	184–185	UV, IR, MS, ^{1}H-NMR (60 MHz)	(57)
Diacetyldihydroisoiguesterin (59)	Iguesterin (14)	$C_{32}H_{42}O_4$ (490.31)	–	165–168	UV, IR, MS, ^{1}H-NMR (90 MHz)	(136)

Table 11 (continued)

Trivial name(s)	R_1	R_2	Derived from	Molecular formula (MW)	$(\alpha)_D$	mp (°C)	Spectral data reported	References
Diacetyldihydro-isotingenone-III (60)	Ac	O	Tingenone (12)	$C_{32}H_{42}O_5$ (506.30)	–	202–204	IR, MS	(21)
Dimethyldihydro-isotingenol-III (61)	Me	H, OH	Tingenone (12)	$C_{60}H_{44}O_3$ (452.33)	–	251–254	IR, MS, ^{1}H-NMR (60 MHz)	(21)
Triacetyldihydro-isotingenol-III (62)	Ac	H, OAc	Tingenone (12)	$C_{32}H_{46}O_6$ (534.33)	–	249–252	IR, MS, ^{1}H-NMR (60 MHz)	(21)

Table 11 (*continued*)

Trivial name(s)	Derived from	Molecular formula (MW)	$(\alpha)_D$	mp (°C)	Spectral data reported	References
Dimethyldihydroiso-iguesterin-III (**63**)	Tingenone (**12**)	$C_{30}H_{42}O_2$ (434.32)	–	177–179	MS, ^{1}H-NMR (100 MHz)	(*21*)

Table 11 (continued)

Trivial name(s)	R_1	R_2	R_3	R_4	Derived from	Molecular formula (MW)	$(\alpha)_D$	mp (°C)	Spectral data reported	References
Diacetylisotingenone-III (64)	Ac	O	Me	H	Tingenone (12)	$C_{32}H_{40}O_5$ (504.29)	—	207–210	UV, IR, MS, ^{1}H NMR (60 MHz)	(21)
Dimethylisotingenone-III (65)	Me	O	Me	H	Tingenone (12)	$C_{30}H_{40}O_3$ (448.30)	—	173–176	IR, MS, ^{1}H-NMR (60 MHz)	(21)
Dimethyl-21β-hydroxy-isopristimerin-III (66)	Me	β-OH, α-H	Me	CO_2Me	21β-Hydroxy-pristimerin (4)	$C_{32}H_{44}O_5$ (508.32)	—	200–203	UV, IR, ^{1}H-NMR (60 MHz)	(24)
Dimethylisocelastrol-III (67)	Me	H_2	Me	CO_2H	Pristimerin (11)	$C_{31}H_{42}O_4$ (478.31)	—	253–254	UV, IR	(95)
Dimethylisopristimerin-III (68)	Me	H_2	Me	CO_2Me	Pristimerin (11)	$C_{32}H_{44}O_4$ (492.32)	—	197–198	UV, IR	(75, 95)
Dimethyl-23-oxoiso-pristimerin-III (69)	Me	H_2	CHO	CO_2Me	23-Oxoiso-pristimerin-III (29)	$C_{32}H_{42}O_5$ (506.30)	—	178–181	IR, ^{1}H-NMR (60 MHz)	(41)

Table 11 (continued)

Trivial name(s)	R_1	R_2	Derived from	Molecular formula (MW)	$(\alpha)_D$	mp (°C)	Spectral data reported	References
Dimethyl-23-nor-6-oxo-pristimerol (70)	H	CO_2Me	23-Nor-6-oxo-pristimerol (33)	$C_{31}H_{42}O_5$ (494.30)	–	semisolid	IR, MS, ^{1}H-NMR (500 MHz), ^{13}C-NMR (75 MHz)	(43)
Dimethyl-6-oxopristimerol (71)	Me	CO_2Me	Pristimerin (11)	$C_{32}H_{44}O_5$ (508.32)	– 79.4° (CHCl$_3$)	217–218	–	(73, 77, 78)
Dimethylzeylasteral (72)	CHO	CO_2Me	Zeylasteral (34)	$C_{32}H_{42}O_6$ (522.30)	– 120.27° (CHCl$_3$)	201–202	UV, IR, ^{1}H-NMR (60 MHz)	(41)
Trimethylzeylasterone (73)	CO_2Me	CO_2Me	Pristimerin (11) Zeylasterone (35)	$C_{33}H_{44}O_7$ (552.31)	– 106.9° (CHCl$_3$)	229–230	UV, IR, MS, ^{1}H-NMR (60 MHz)	(77, 78)

have been encountered as natural products. Therefore, it was considered desirable to tabulate their known data and origin. Data available for some important derivatives of celastroloids are listed in Table 11.

6. The Chemistry of Celastroloids

6.1. Isolation Techniques

GISVOLD, who was responsible for conducting the original scientific studies on *Celastrus paniculatus* was interested in the chief pigment found in the outer root bark of this plant (*47*). His isolation technique involving fractional extraction of the freshly collected plant material with Skelly-solve B was aimed at obtaining the major pigment, celastrol (**1**). The pigment which separated during the process of extraction was purified by recrystallization from a mixture of isopropyl ether and Skelly-solve B, thus furnishing pure celastrol.

The selection of plant parts for the extraction of the antibiotic principle occurring in *Pristimeria indica* was guided by biological (anti-bacterial) activity (*6*). The yellow outer bark (phellum) of the root was found to be more active than the inner red bark and the pith. Extraction of the phellum with petroleum ether followed by evaporation of the solvent afforded a gummy, dark orange-colored mass. This was dissolved in a minimum quantity of boiling sulfuric ether (diethyl ether) and left overnight in a refrigerator to yield an orange-colored crystalline substance. A two step sequential recrystallization from ethyl acetate followed by a mixture of benzene and petroleum ether afforded pure pristimerin (**11**). This straightforward process of pristimerin isolation has been patented (*8*). Tingenone (**12**), the major pigment present in the stem bark of *Euonymus tingens*, was isolated in an analogous fashion (*9*). The main difference however, was that after precipitation of tingenone the mother liquor was subjected to silica gel column chromatography to yield 20α-hydroxytingenone (*6*).

Although partial purification of the major celastroloids can be achieved by precipitation, the isolation of related minor constituents co-occuring with other triterpenoid types relies heavily on chromatographic techniques. Both silica gel column chromatography and preparative TLC have been employed in the separation and purification of minor celastroloids. The use of silica gel appears to be imperative since Al_2O_3, $MgCO_3$ and MgO supports were found to be inappropriate for the purification of pristimerin present in a crude extract. This failure has been explained as probably due to "lake" formation between the pigment and the adsorbent (73).

Separation of two phenolic triterpenoids, isopristimerin-III (27) and isotingenone-III (28), found in *Maytenus ilicifolia* has recently been achieved by a Kupchan type fractionation involving solvent-solvent partitioning of the hot methanolic extract between water and chloroform, and *n*-butanol, successively. The *n*-butanol soluble fraction was subjected to Dianionic HP-20 column chromatography using a water-methanol solvent system and 80% acetone. The 80% acetone eluate was then concentrated and partitioned between water and ethyl acetate. Final purification of the ethyl acetate soluble fraction utilizing reversed phase and Sephadex LH-20 column chromatography yielded the desired constituents (93). Purification of pristimerin in *Celastrus dispermus* and *Denhamia pittosporoides* was achieved by countercurrent distribution (73). Preliminary determination of the distribution coefficients of the crude extract suggested cyclohexane-95% aqueous methanol as a suitable system. Moreover, the addition of NaCl has been found to suppress emulsification and to increase the settling rate. In a countercurrent distribution of the extract (4 g) contained in three tubes, 98 transfers have been carried out. The contents of each tube were evaporated to dryness under reduced pressure and the organic materials separated from NaCl by ether extraction. The ethereal extracts were evaporated to dryness, weighed and a graph of weight versus tube number was constructed. Peaks appeared at tube numbers 3, 6–8, 39–42, 69–71, and 87–92. Tubes 65–76 inclusive contained a diterpene, whereas the contents of tubes 31–48 inclusive gave orange needles (150 mg, 0.37%) of pristimerin (11) when crystallized (73).

In an investigation of the pristimerin-like triterpenoids of the stem bark of *Euonymus tingens*, NAKANISHI, GOVINDACHARI and coworkers (120) subjected a hexane extract to silica gel chromatography giving a TLC pure fraction which on evaporation afforded orange crystals, m.p. 155–165°. HPLC analysis [Waters ALC–100, 3 × 3 ft. Corasil II, hexane-chloroform-acetonitrile (10:1:0.5)] of the "crystals" indicated the presence of at least 11 components. Fractionation of 560 mg of original "crystals" afforded two major constituents, tingenin A [= tingenone (12)] (137 mg) and tingenin B [= 22β-hydroxytingenone (8)] (36 mg). Attempted separation of balaenol (15) and isobalaenol (18) by HPLC has been reported to be unsuccessful (37).

6.2. Structure Elucidation

6.2.1. Early Structural Studies of Celastrol and Pristimerin

As illustrated in the following discussion, elucidation of the currently accepted structures of celastrol and pristimerin took nearly two and a half decades of intense research by several groups of natural products chemists from England, India, Japan, Scotland and the USA. Although anecdotal, it is interesting to note that the leader of the English team had once stated that the tree from which pristimerin was obtained was called the "Ph.D. tree" because the studies of his group had led to several Ph.D. degrees (121).

Structural studies on the red pigment of *Celastrus scandens* were initiated by GISVOLD in 1939 with the aim of determining whether it was identical with β-carotene (47) and who concluded that the pigment was different from β-carotene. During the period 1939–1942 GISVOLD published a series of papers on the structure of celastrol (47–50), the last one of which suggested that celastrol was either a mono- or di-alkyl substituted β- or α-naphthoquinone of one of the following tentative structures (Fig. 3).

In 1942 FIESER and JONES carried out extensive UV spectral studies on celastrol and its methyl derivative (40). Comparison of their data with those of natural naphthoquinone pigments such as plumbagin, phthiocol, hydroxydroserone and vitamins K_1 and K_2 made them support GISVOLD's *ortho*-quinonoid structure for celastrol (Fig. 3) [R_1 = Me and R_2 = a hydrogeranyl or homohydrogeranyl group]. Further evidence for GISVOLD's *peri*-hydroxy-*ortho*-naphthoquinone structure was provided by UV spectral studies conducted by SCHECHTER and HALLER (132).

KULKARNI and SHAH in 1954 initiated structural studies on pristimerin (102) previously isolated by BHATNAGAR and DIVEKAR from the Indian medicinal plant, *Pristimeria indica* (6) and noted the similarity between the UV absorption curves of pristimerin and celastrol. Since the

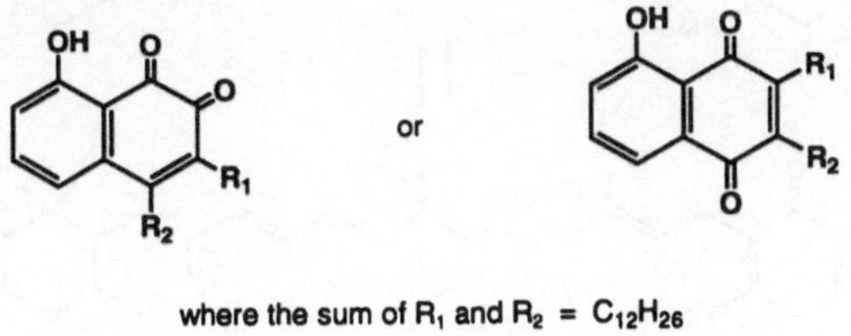

where the sum of R_1 and R_2 = $C_{12}H_{26}$

Fig. 3. GISVOLD's tentative structures for celastrol

m.p. of an admixture of pristimerin and methylated celastrol remained undepressed they concluded that pristimerin and methyl celastrol were probably identical. Based upon IR and UV spectroscopic data for the parent compound as well as its derivatives, a benzoquinone structure discounting GISVOLD's naphthoquinone structure was proposed. Preliminary selenium dehydrogenation experiments had indicated that pristimerin may be a steroid or a terpenoid. This observation and the fact that triterpenes have been isolated from the same plant species led KULKARNI and SHAH to advance the partial structure for pristimerin depicted in Fig. 4 (102).

Detailed interpretations of the IR spectrum of pristimerin were carried out by the groups of BHATNAGAR (98) and NAKANISHI (118) in 1955. BHATNAGAR and coworkers also performed X-ray and optical investigations and Patterson projection maps were calculated from the X-ray data which suggested that pristimerin should contain 20–25 atoms of carbon and oxygen, excluding hydrogen. As a result of these chemical, spectral and crystallographic studies, BHATNAGAR's group concluded that pristimerin is a pigment of the oxygenated carotenoid type and proposed a xanthophyll like structure (Fig. 5). Although the formation of a product with UV absorptions similar to tetramethylchrysene was noted in their micro zinc-dust distillation of pristimerin, they incorrectly assumed that

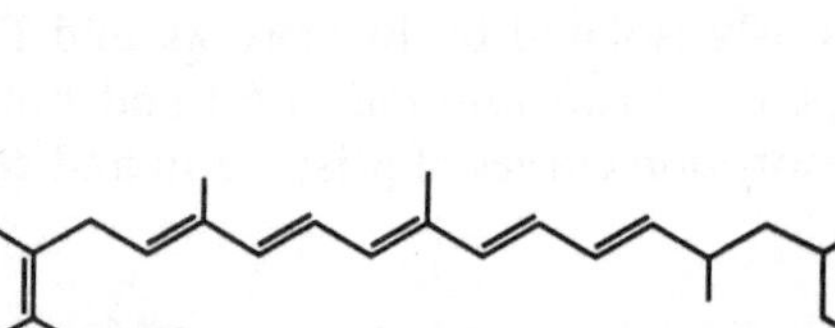

$C_{17}H_{28-30}$ with one double bond

Fig. 4. KULKARNI's proposed structure for pristimerin

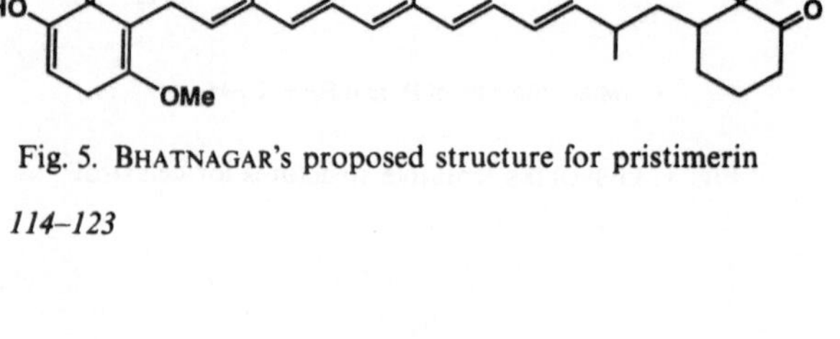

Fig. 5. BHATNAGAR's proposed structure for pristimerin

a cyclization had occurred in the course of this pyrolytic distillation, an assumption which caused them to deviate from the terpenoid/steroid structure proposed by KULKARNI and SHAH (*102*).

Detailed UV and IR spectral data for a series of pristimerin and celastrol derivatives were reported by NAKANISHI's group (*119*). Based on these and numerous chemical reactions, NAKANISHI and coworkers advanced a pentacyclic *ortho*-quinonoid structure for pristimerin (see Fig. 6). A 5-membered ring C was proposed to explain the IR data, although a 6-membered ring C was suggested as an alternative. In 1957 GRANT and JOHNSON published two papers (*74, 75*) dealing with the nature of the chromophore of pristimerin. The correct molecular formula for pristimerin, $C_{30}H_{40}O_4$, was suggested based on the results of repeated microanalysis. Molecular formulae of some reaction products together with their chemical and spectroscopic (IR and UV) properties allowed these authors to propose partial structures for pristimerin and celastrol (Fig. 7).

In part II of their series of publications on the structure of pristimerin, SESHADRI, KULKARNI and coworkers (*134*) presented chemical evidence to support their previous structure (Fig. 4), thus discounting the structures subsequently postulated by NAKANISHI (*119*) and GRANT and JOHNSON (*74, 75*). COOKE and THOMSON who reviewed naturally occurring

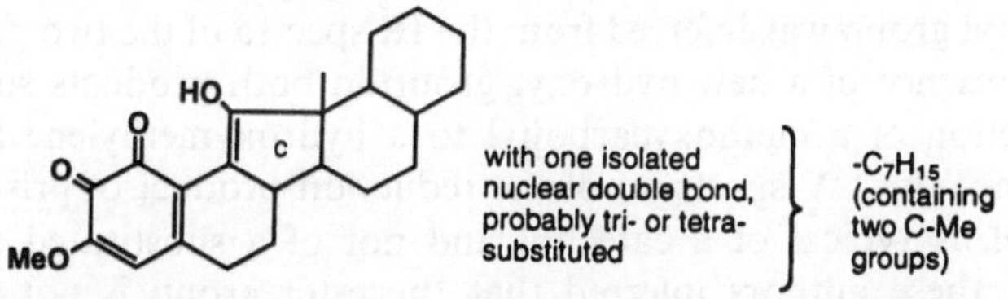

Fig. 6. NAKANISHI's first proposed structure for pristimerin

Pristimerin   R = OMe or Me
Celastrol   R = OH   or Me

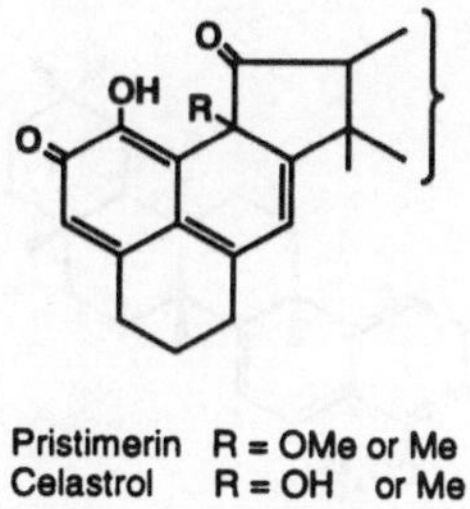

Fig. 7. GRANT and JOHNSON's proposed structures for celastrol and pristimerin

quinonemethides and related compounds in 1958 (*16*) considered the formation of a picene derivative during the zinc-dust distillation to be a significant result and suggested the possibility of a relationship between pristimerin and the pentacyclic triterpenes. They also stated that "a modified friedelane skeleton is most attractive because known compounds (e.g., cerin) have the required distribution of groups for the quinone methine chromophore now postulated, and because *Siphonodon australe* Benth., the richest source of friedelane derivatives, is a plant of the family Celastraceae." The structure proposed for pristimerin by COOKE and THOMSON (Fig. 8), although not supported at the time by experimental evidence, may be considered as a landmark in the history of pristimerin and celastrol.

In 1960 JOHNSON and coworkers reexamined the structure (Fig. 7) proposed by them earlier. Investigation of some reaction products suggested that the "inert carbonyl" group of pristimerin is part of an ester function, and is not conjugated with the main chromophore. Moreover, they found that the bottom ring, postulated to account for the existence of different naphthalenoid acid rearrangement products of pristimerin, was no longer necessary. Contrary to the claim by SESHADRI, KULKARNI and coworkers (*134*), Zeisel determination of the LiAlH$_4$ reduction products of two derivatives of pristimerin, namely pristimerol dimethyl ether and the naphthalenoid dimethyl ether derived from the Thiele acetylation product, showed that one methoxy group was lost. Loss of the carbonyl group was inferred from the IR spectra of the two products, and the presence of a new hydroxyl group in both products suggested that reduction of a methoxycarbonyl to a hydroxymethylene had occurred. Since the UV spectrum of the reduction product of pristimerin, pristimerol, is typical of a catechol and not of a substituted salicylic acid ester, these authors inferred that the ester group is not directly attached to the chromophore as suggested by COOKE and THOMSON (*16*). Although JOHNSON and coworkers implied that the methoxycarbonyl group was present in rings C, D or E of the pentacyclic triterpenoid

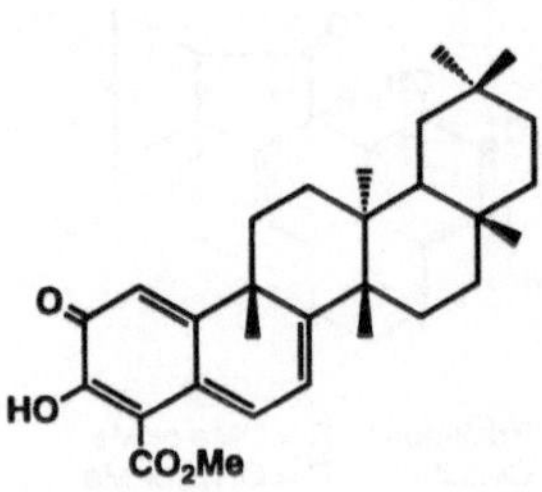

Fig. 8. COOKE and THOMSON's proposed structure for pristimerin

skeleton, they were unable to determine its exact location. The possible structures proposed by JOHNSON *et al.* are shown in Fig. 9.

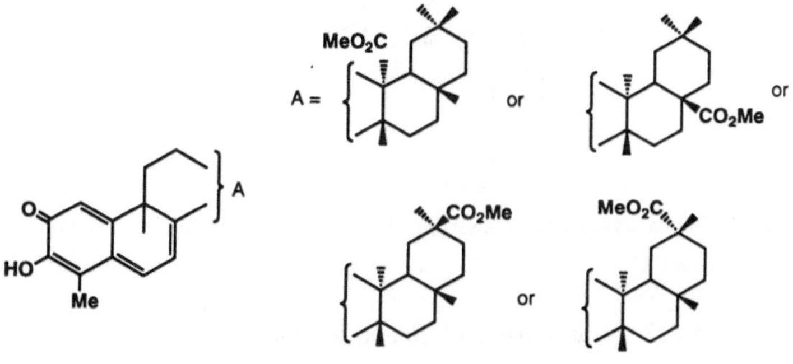

Fig. 9. Structures for pristimerin proposed by JOHNSON and coworkers

Application of ^{1}H-NMR (30 and 60 MHz) and mass spectroscopy aided NAKANISHI's group in extending the partial structure proposed for pristimerin by JOHNSON and coworkers to the full expression represented in Fig. 10 (*88, 118*). In the ^{1}H-NMR spectra of pristimerin and several of its derivatives, signals due to methyl groups appeared as singlets and this suggested an oleanane type triterpene carbon framework for pristimerin. Furthermore, the ^{1}H-NMR and MS data for these compounds supported the chromophore-bearing A, B ring system of pristimerin suggested by GRANT and coworkers. The position of attachment of the methoxycarbonyl group in the oleanane triterpene skeleton was determined by judicious application of chemical reactions and ^{1}H-NMR spectral analysis of the reductive triacetate (74) obtained by the treatment

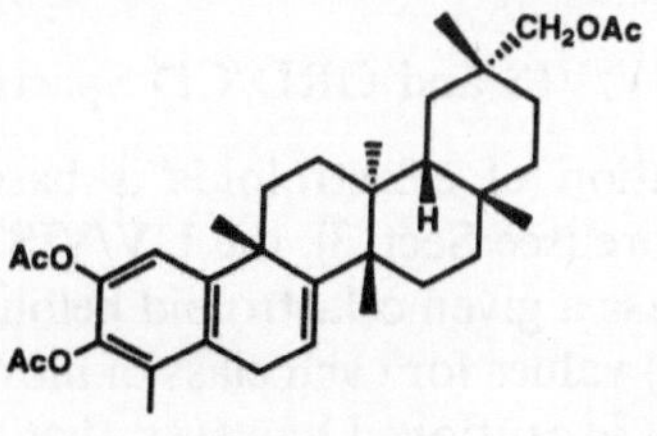

74

of pristimerin with LiAlH$_4$ followed by acetylation (*88*). The final paper describing the currently accepted structure of pristimerin was published in 1963 by JOHNSON and coworkers (*95*) who provided further evidence for the structure (Fig. 10) proposed by NAKANISHI's group (*88*).

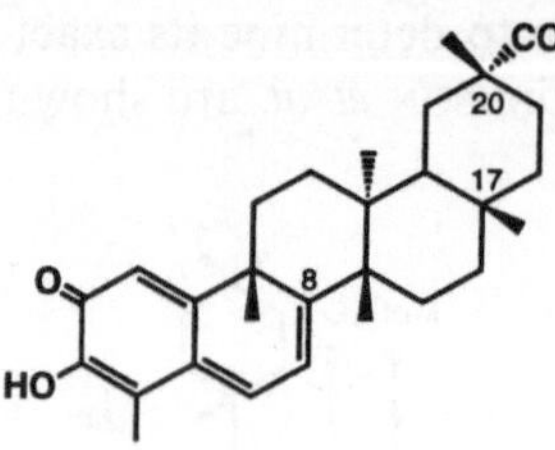

R = H   Celastrol   (1)
R = Me   Pristimerin   (11)

Fig. 10. NAKANISHI's and JOHNSON's structures for celastrol and pristimerin

6.2.2. Application of Spectroscopic Techniques

Prior to the advent of nuclear magnetic resonance (NMR) techniques, the structure investigation of celastroloids depended largely on chemical transformations coupled with the application ultraviolet/visible (UV/VIS) and infrared (IR) spectroscopy. In the absence of NMR and mass spectrometric (MS) techniques, the types of tedious chemical transformations which were outlined in the previous section would have continued for several more years.

The spectroscopic techniques currently employed in structure elucidation of celastroloids include UV/VIS, optical rotatory dispersion/circular dichroism (ORD/CD), IR, ^{1}H-NMR, ^{13}C-NMR spectroscopy, MS and X-ray crystallography. Application of each of these techniques in solving structural problems will be discussed. Where relevant, some typical spectra are given since the overall features of a spectrum can be seen more effectively from an illustration than from numerical data in a table or text.

6.2.2.1. UV/VIS and ORD/CD Spectroscopy

Since the classification of celastroloids is based primarily on the constituent chromophore (see Sect. 3), the UV/VIS spectra are useful in identifying to which class a given celastroloid belongs. The approximate UV/VIS λ_{max} (and ε_{max}) values for each class of monomeric celastroloids are listed in Table 12. It is cautioned however, that any functional group directly attached to the chromophore may cause some deviation from these standard values.

The UV/VIS spectrum of pristimerin (11), for example, is quite characteristic (Fig. 11). The spectrum reveals two prominent absorption bands of almost equal intensity at 425 and 255 nm, with a shoulder at

Table 12. *Approximate UV/VIS Spectral Data for Monomeric Celastroloid Classes*

Class	Approx. λ_{max} (ε_{max}) nm			
Quinonemethide	255 (3.90)			425 (4.00)
14(15)-Enequinonememethide	255 (3.90)			445 (4.00)
9(11)-Enequinonemethide	255 (4.00)			445 (4.00)
Phenolic	255 (4.50)	290–310 (4.00)		445 (4.00)
6-Oxophenolic	245–255 (4.00)	290–305 (3.70)	310–340 (3.70)	
7-Oxoquinonemethides	245 (3.50)	320 (3.90)	330 (3.90)	410 (3.50)

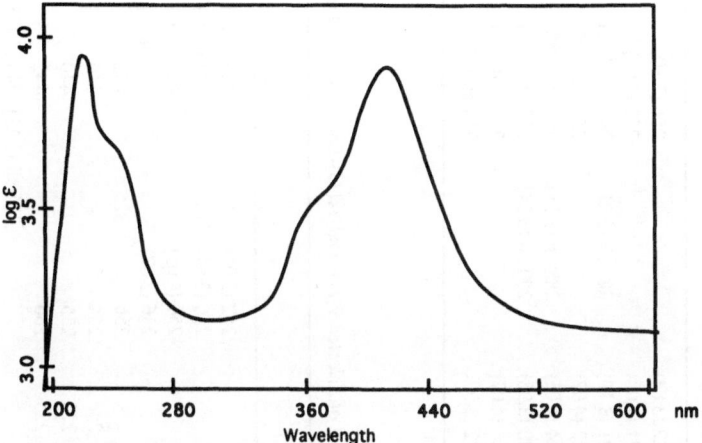

Fig. 11. UV/VIS spectrum of pristimerin (**11**) in EtOH

335–340 nm. These two main bands appear to be characteristic of this class of celastroloids (Table 13). In fact, it was the similarity in UV absorption curves of celastrol and pristimerin which made KULKARNI and SHAH to suspect their structural interrelationship (*102*). Although the pH dependence of the UV/VIS spectra of celastroloids has not routinely been used as a tool in structural studies, the presence of only one set of isosbestic points in the UV/VIS spectrum of pristimerin measured in alcoholic alkali [λ_{max} 362 (ε_{max} 4800) and 495 (ε_{max} 3300) nm] has been used to draw the conclusion that it exists solely in the enol-enolate form (*118*). It has also been observed that the longer wavelength band at λ_{max} 425 nm undergoes a bathochromic shift to ca. 525 nm on addition to alkali. The UV/VIS spectral data for some representative 9(11)- and

Table 13. *UV/VIS Spectral Data for Some Quinonemethide Triterpenoids*

Compound	λ_{max} (log ε)	Solvent	Reference(s)
Celastrol (1)	422 (3.84) 256 (3.80)	EtOH	(40, 122)
21β-Hydroxypristimerin (4)	421 (4.08) 250 (3.94)	EtOH	(24)
20-Hydroxy-20-*epi*-tingenone (7)	421 (3.10) 288 sh (2.21) 242 sh (2.98)	EtOH	(122)
22β-Hydroxytingenone (8)	422 (4.02) 253 sh (3.30)	MeOH	(3)
Pristimerin (11)	424 (4.01) 288 sh (3.28) 254 sh (3.88)	EtOH	(57, 120)
Tingenone (12)	420 (3.98) 290 sh (3.34) 253 sh (3.87)	MeOH	(21, 57, 120)
Iguesterin (13)	420 (4.07) 254 sh (4.02)	EtOH	(57)
Isoiguesterin (14)	422 (3.97) 254 sh (3.80)	EtOH	(136)

Table 14. *UV/VIS Spectral Data of Selected 9(11)- and 14(15)-Enequinonemethide Triterpenoids*

Compound	λ_{max} (log ε)	Solvent	Reference(s)
Balaenol (15)	252 (3.09) 446 (4.14)	EtOH	(37)
Balaenonol (16)	256 (3.94) 444 (4.02)	EtOH	(37)
Hydroxypristimerinene (24)	256 (4.05) 446 (4.09)	MeOH	(24)
Isobalaenediol (17)	256 (3.02) 441 (4.01)	EtOH	(37)
Netzahualcoyene (19)[a]	230 256 325 442	EtOH	(65)
Netzahualcoyol (20)[a]	215 243 254 440	EtOH	(65)
Netzahualcoyondiol (21)[a]	203 228 255 440	EtOH	(65)
Netzahualcoyonol (23)[a]	214 235 260 442	EtOH	(65)
Salacia quinonemethide (26)	325 sh (3.11) 448 (4.17)	MeOH	(129, 130)

[a] Absorption maxima not reported.

Table 15. *UV/VIS Spectral Data of Some Phenolic Triterpenoids*

Compound	λ_{max} (log ε)	Solvent	Reference(s)
Diacetylisotingenone-III (64)	250 (4.90), 290 (4.04)	EtOH	(21)
Dimethyl-21β-hydroxy-isopristimerin-III (66)	254 (4.50), 298 (3.90)	MeOH	(24)
Dimethylisocelastrol-III (67)	249 (4.59), 255 (4.58), 283 sh (3.69), 301 (3.81)	EtOH	(95)
Isopristimerin-III (27)	252 (4.30), 256 (4.32), 305 (3.52), 376 (3.51)	EtOH	(93)
Isotingenone-III (28)	251 (4.39), 256 (4.40), 308 (3.55), 376 (3.34)	EtOH	(93)
23-Oxoisopristimerin-III (29)	263 (3.15), 277 sh (3.17)	EtOH	(41)

Table 16. *UV/VIS Spectral Data of Some 6-Oxophenolic Triterpenoids*[a]

Compound	λ_{max} (log ε)	Reference(s)
Demethylzeylasteral (27)	207 (3.31), 247 (3.27), 304 (2.85), 388 (2.82)	(41)
Demethylzeylasterone (31)	210 (4.15), 250 (3.99), 302 (3.67), 340 (3.64)	(41)
Dimethyl-6-oxopristimerol (71)	210 (4.13), 225 (4.00), 247 (4.07), 285 (3.80), 300 (3.92)	(78)
Dimethylzeylasteral (72)	211 (4.32), 222 (4.29), 248 (4.20), 296 (3.91), 315 (3.97)	(41)
Trimethylzeylasterone (73)	207 (4.00), 225 (3.88), 245 (3.99), 287 (3.72), 312 (3.72)	(78)
Zeylasteral (34)	207 (3.99), 247 (3.94), 306 (3.47), 399 (3.49)	(41)
Zeylasterone (35)	211 (4.19), 226 (4.05), 255 (4.08), 295 (3.79), 340 (3.70)	(41, 78)
23-*Nor*-6-oxodemethylpristimerol (32)	203 (4.29), 226 (3.96), 254 (4.05), 298 (3.70), 335 (3.71)	(43)

[a] All UV/VIS spectra recorded in EtOH.

14(15)-enequinonemethide, phenolic and 6-oxophenolic triterpenoids are compiled in Tables 14, 15, and 16, respectively. The H_3BO_3-NaOAc induced shifts in the UV spectra have been used to determine the presence of an *ortho*-dihydroxy system in 6-oxophenolic triterpenoids (*78*).

Optical rotatory dispersion (ORD) and circular dichroism (CD) spectra were occasionally used to obtain stereochemical information on celastroloids. During the structure elucidation of the 7-oxoquinonemethide triterpenoid dispermoquinone (**36**), MARTIN suspected the α-axial stereochemistry for the C-8 proton of the tetrol (**91**) (see Scheme 10; Sect. 6.3.3) (*112*). This stereochemical assignment was confirmed by comparing the ORD curve of dispermoquinone with that of maytenoquinone, a quinonemethide diterpenoid with known stereochemistry. The ORD curves of the two compounds were found to be mirror images suggesting that these were optical antipodes. The CD curves of pristimerin (**11**), zeylasterone (**35**), balaenol (**15**) and balaenonol (**16**) have been recorded and in some cases used to obtain stereochemical information (*37, 78*). The CD spectra of pristimerin (**11**) and balaenol (**15**) representing two typical examples are reproduced in Fig. 12. It is note-

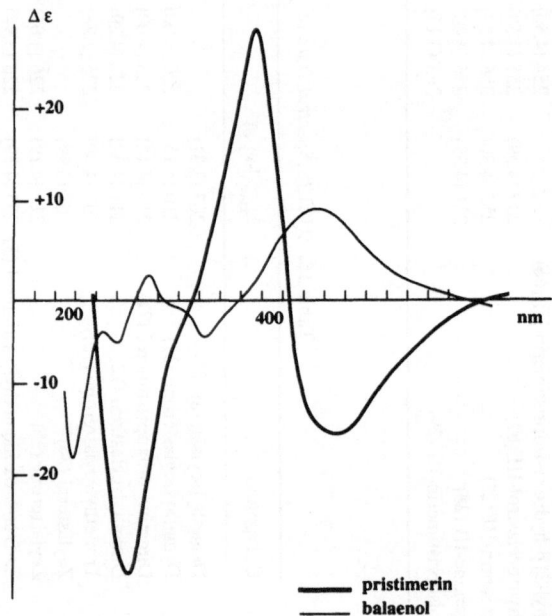

Fig. 12. CD spectra of pristimerin (**11**) and balaenol (**15**) in CH_3CN

worthy that changes in CD and UV spectra of tingenone (**12**) have been used to study its interaction with DNA (*12*).

6.2.2.2. Infrared Spectroscopy

IR spectroscopy provided both helpful and in some cases, misleading information during the early structural studies on celastrol and pristimerin (*73, 117, 118*). Only after the correct structures for these celastroloids and their derivatives were established was it possible to interpret the IR spectra of these compounds. For a discussion of the interpretation of the unusual carbonyl absorption frequency of isopristimerin-I, see Sect. 6.3.5.

The IR spectrum of pristimerin in CCl_4 is reproduced in Fig. 13. As expected, it shows bands for the chelated hydroxy (3380 cm^{-1}), ester carbonyl (1740 cm^{-1}) and the conjugated carbonyl of the quinonemethide system (1607 cm^{-1}). Enequinonemethides also show a band at about the same frequency for the enequinonemethide carbonyl and thus IR spectroscopy cannot be used to differentiate between these two classes. IR spectra of 6-oxophenolic triterpenoids, depending on the substituent at C-4, show a characteristic band at 1620–1640 cm^{-1} for the conjugated C-6 carbonyl group. In 7-oxoquinonemethide triterpenoids bands due to both quinonoid carbonyls are present, the H-bonded carbonyl at C-2 absorbing around 1620–1625 cm^{-1} and the remaining carbonyl appearing around 1660–1670 cm^{-1}. The similarity between the IR spectra of dispermoquinone (**36**) and maytenoquinone in the region 1440–1700 cm^{-1} has been used to infer the presence of a 7-oxoquinonemethide

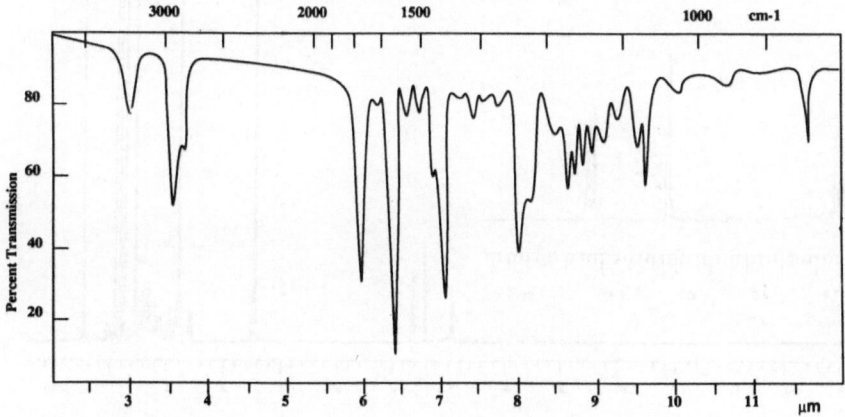

Fig. 13. IR spectrum of pristimerin (**11**) in CCl_4

system in the former (*112*). IR spectra of celastroloids are useful in determining the presence or absence of certain functional groups outside the A and B rings especially in ring E.

6.2.2.3. NMR Spectroscopy

During the last two decades NMR spectroscopy has been extensively utilized for structure elucidation of celastroloids, leading to the revision of structures in a few select instances. This is illustrated in the following applications of 1D [¹H-NMR, ¹³C-NMR, nOe (nuclear Overhauser effect), nOe difference] and 2D [¹H-¹H COSY (correlation spectroscopy), HETCOR (heteronuclear correlation spectroscopy), HMQC (heteronuclear multiple-quatnum correlation), HMBC (heteronuclear multiple bond correlation), Selective INEPT (insensitive nuclei enhanced by polarization transfer), ROESY (rotating frame nuclear Overhauser spectroscopy)] techniques. The use of sophisticated 2D NMR techniques has also resulted in the revision of spectral assignments for some celastroloids previously made using chemical shift arguments. High temperature (150 °C) NMR studies in dimethyl sulfoxide have been used to demonstrate the interconversion of the dimeric celastroloids, cangorosin

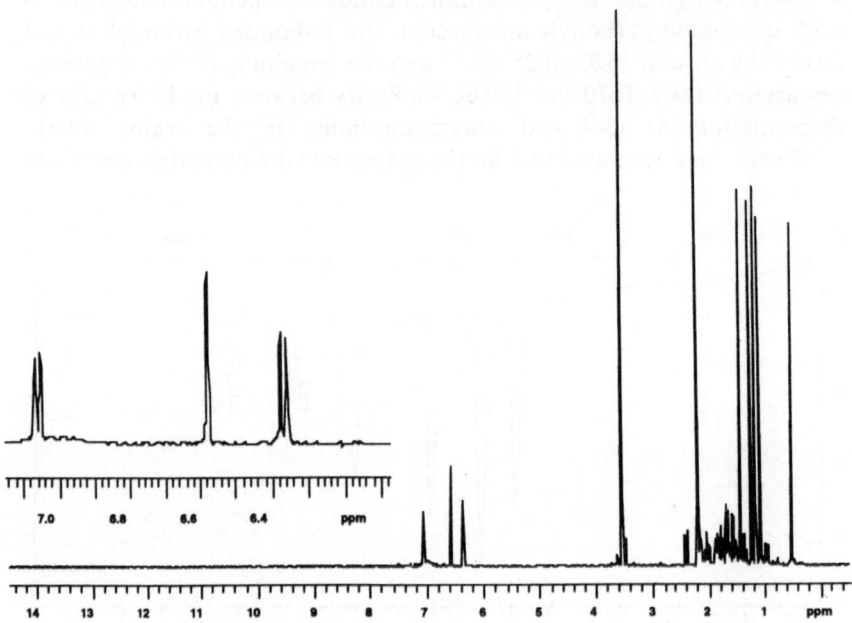

Fig. 14. ¹H-NMR spectrum (400 MHz) of pristimerin (11) in CDCl₃

A (**38**) and its atropisomer (**39**) and to estimate the half-life (3 hr) and the rotational barrier (32.8 kcal mole^{-1}) for this conversion (*92*).

6.2.2.3.1 ^{1}H-NMR Spectroscopy. Despite two decades of derivatization, degradation, and UV and IR studies, the correct structure of pristimerin had to await the application of ^{1}H-NMR spectroscopy, albeit at 30 MHz (*88, 121*). The ^{1}H-NMR spectra of celastroloids are characterized by the presence of two main groups of resonances, one in the olefinic/aromatic region and the other in the aliphatic region. The ^{1}H-NMR spectrum (400 MHz) of pristimerin (**11**) reproduced in Fig. 14, also contains resonances due to carbomethoxy and vinyl methyl groups.

Assignments of olefinic, carboxymethoxy and vinyl methyl protons of pristimerin and several related quinonemethides were made by early workers using chemical shifts and coupling constant arguments (*88, 97, 120*). However, the assignment of resonances in the region δ 0.5–2.5 ppm due to the aliphatic methines, methylene and non-vinylic methyl groups of pristimerin (**11**) required more rigorous analysis and was achieved by 2D ^{1}H-^{1}H and ^{1}H-^{13}C COSY experiments (*81*). Coupling constants for these protons were obtained by careful analysis of the fine structures of the cross-peaks observed in the ^{1}H-^{1}H COSY spectrum of this region (Fig. 15). Application of these 2D NMR techniques has also aided to revise the assignments previously made (*78*) for the signals due to 13-CH$_3$, 17-CH$_3$, and 20β-CH$_3$. ^{1}H-NMR spectral data of selected quinonemethide triterpenoids are presented in Tables 17–19. The data in these and other tables in this section were extracted directly from cited literature references and therefore the assignments have varying degrees of accuracy.

The separation of overlapping methyl singlets in the ^{1}H-NMR (60 MHz) spectrum of maitenin [tingenone (**12**)] in a 1:1 mixture of CDCl$_3$:C$_6$H$_6$ has been observed by MARINI-BETTOLO's group (*18*). In an attempt to determine the stereochemistry at C-20 of 20-hydroxytingenone by application of NMR techniques, CORDELL and his coworkers found that the overlapping ^{1}H resonances of C-26-Me and C-30-Me in CDCl$_3$ could be separated using a mixture of CDCl$_3$:C$_6$D$_6$ (1:1) (*107*). Definitive assignment of these two methyl signals were then made through the analysis of 2D NMR spectra such as ^{1}H-^{1}H COSY, ROESY, HMQC, HMBC and selective INEPT. These studies helped to assign the configuration of C-20 in 20-hydroxytingenone as 20β-hydroxy-20α-methyl suggesting that the structure previously proposed for 20α-hydroxytingenone (**6**) should be revised to 20-hydroxy-20-*epi*-tingenone (**7**). The revised stereochemistry at C-20 was further supported by nOe studies (*107*).

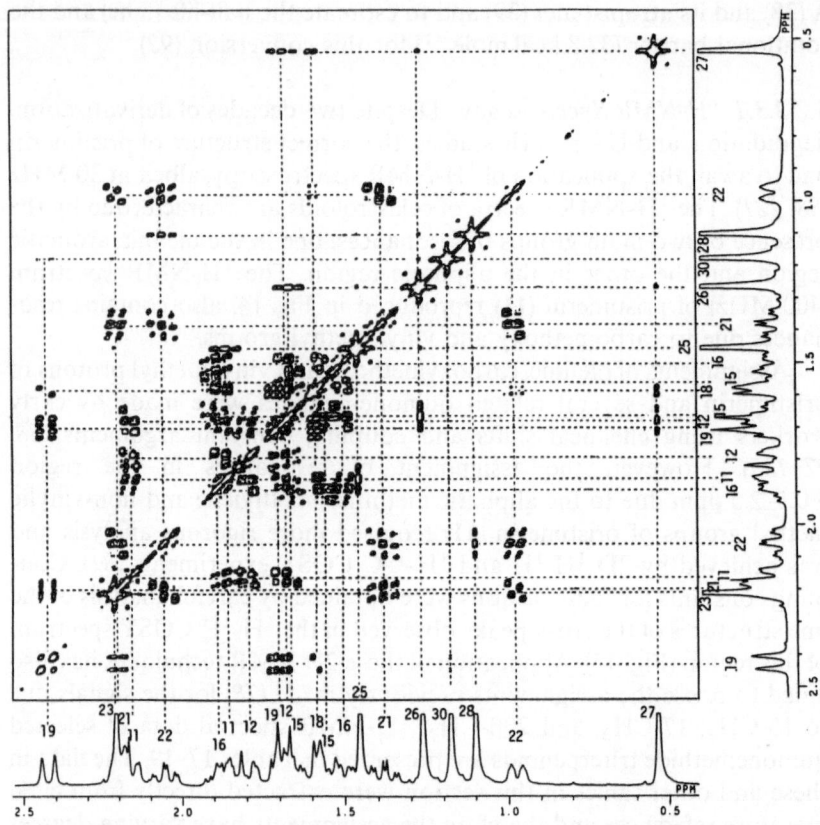

Fig. 15. Contour plot of the high-field region of the ¹H-¹H COSY spectrum of pristimerin
(11) in CDCl₃

A recent report described the application of 2D ¹H-¹H and ¹H-¹³C
shift correlation NMR and nOe techniques to complete assignments of
the ¹H-NMR spectra of three 14(15)-enequinonemethide triterpenoids,
balaenol (15), balaenonol (16) and isobalaendiol (17) (140). Coupling
constants obtained in this study have suggested half-chair/chair conformations for D/E rings of balaenonol which was supported by nOe
experiments. The observed nOe enhancement of 7-H on irradiation of
the vinyl methyl signal (at δ 1.77 ppm) supported the 14(15)-enequinone-
methide structure proposed for balaenonol (16) rather than the previously proposed 9(11)-enequinonemethide structure (37). ¹H-NMR data
for 3 selected 14(15)-enequinonemethides for which complete assignments have been made are given in Table 20.

Table 17. ^1H NMR Spectral Data of Selected Quinonemethide Triterpenoids[a,b]

Protons on	Celastrol (1) (122)	Excelsine (3) (11)	21β-Hydroxy-pristimerin (4) (24)	20-Hydroxy-20-epi-tingenone (7) (122)	Iguesterin (13) (57)
C-1	6.47 (br s)	6.43 (d, 1.5)	6.46 (d, 1.5)	6.53 (br s)	6.56 (s)
C-6	7.04 (br d, 6.8)	6.95 (dd, 8, 1.5)	6.96 (dd, 7, 1.5)	7.02 (br d, 5.3)	7.03 (d, 7)
C-7	6.30 (d, 7.1)	6.25 (d, 8)	6.23 (d, 7)	6.37 (d, 6.4)	6.34 (d, 7)
C-19	2.48 (d, 16.1)	—	—	2.48	—
C-21	—	4.06 (dd, 11.5, 4.5)	4.00 (dd, 11.5, 4.5)	—	5.24 (m)
C-21 OH	—	5.25 (br s)	5.30 (s)	—	—
C-22	0.90 (br d, 13.7)	—	—	2.99 (d, 14.2)	—
C-23-Me	2.18 (s)	2.22 (s)	2.20 (s)	2.22 (s)	2.31 (s)
C-25-Me	1.40 (s)	1.46 (s)[c]	2.42 (s)[c]	1.46 (s)	1.48 (s)[c]
C-26-Me	1.25 (s)	1.25 (s)[c]	1.23 (s)[c]	1.35 (s)[d]	1.32 (s)[c]
C-27-Me	0.54 (s)	0.83 (s)[c]	1.22 (s)[c]	0.88 (s)	0.51 (s)[c]
C-28-Me	1.06 (s)	1.03 (s)[c]	1.03 (s)[c]	1.12 (s)	1.00 (s)[c]
C-29-Me	—	—	0.47 (s)[c]	—	1.61 (d, 2)
C-30-Me	1.21 (s)	1.17 (s)	—	1.35 (s)[d]	—
C-29-CH$_2$OH	—	3.56 and 3.26 (each d, 10)	—	—	—
OCH$_3$	—	—	3.53 (s)	—	—
C-30-CH$_2$	—	—	—	—	4.58 (br s)

[a] Measured in CDCl$_3$. Chemical shifts (δ) are expressed in parts per million from Me$_4$Si and coupling constants (J) in Hertz.
[b] Multiplicity: s = singlet, d = doublet, m = multiplet.
[c] Not assigned to any particular methyl group.
[d] Overlapping signals.

Table 18. ¹H NMR Spectral Data of Selected Quinonemethides[a,b]

Protons on	15α,22β-Dihydroxy-tingenone[c] (2) (28)	30-Hydroxypristimerin[d] (5) (27)	20α-Hydroxytingenone[d] (6) (81)	22β-Hydroxytingenone[c] (8) (28)
C-1	6.80 (d, 1)	6.52 (d, 1.4)	6.53 (d, 1)	6.78 (d, 1)
C-3 OH	6.09 (s)	6.98 (br s)	6.97 (s)	–
C-6	6.91 (dd, 7, 1)	7.01 (dd, 7, 1.4)	7.01 (dd, 7, 1)	6.96 (dd, 7, 1)
C-7	6.84 (d, 7)	6.35 (d, 7)	6.36 (d, 7)	6.24 (d, 7)
C-11α	2.07 (m)	1.86 (td, 14, 5)	1.95 (td, 14, 7)	1.90 (td, 13, 7)
C-11β	–	2.16 (ddd, 14, 4.5, 2)	2.21 (ddd, 14, 5, 3)	2.14 (ddd, 13, 5, 2.5)
C-12α	–	1.76 (ddd, 14, 5, 2)	1.76 (ddd, 14, 7, 3)	–
C-12β	1.70 (m)	1.68 (td, 14, 14.5)	1.82 (td, 14, 5)	1.67 (m)
C-15α	–	1.68 (td, 13.5, 5)	1.84 (ddd, 12.5, 10.5, 4.5)	1.79 (td, 15, 4.5)
C-15β	4.54 (br s)	1.58 (ddd, 13.5, 6, 2)	1.73 (m)	1.51 (ddd, 15, 9, 2.5)
C-16α	2.92 (dd, 15.5, 2)	1.52 (ddd, 13.5, 5, 2)	1.65 (ddd, 13.5, 4.5, 2.5)	2.56 (br dd, 15, 4.5)
C-16β	1.89 (dd, 15.5, 3.5)	1.90 (td, 13.5, 6)	1.91 (ddd, 13.5, 10.5, 5)	1.55 (td, 15, 5)
C-18	1.86 (d, 6.5)	1.62 (br d, 8)	1.93 (dd, 9, 5)	1.71 (m)
C-19α	2.25 (dd, 14.5, 6.5)	2.25 (br d, 15)	2.28 (dd, 15, 9)	2.14 (dd, 13, 6.5)
C-19β	1.77 (ddd, 14.5, 13.5, 6.5)	1.69 (dd, 15, 8)	2.20 (dd, 15, 5)	1.71 (m)
C-20α	2.80 (ddq, 13.5, 6.5, 6.5)	–		2.67 (ddq, 13, 6.5, 6.5)
C-20 OH			3.24 (s)	–
C-21α		2.26 (ddd, 14, 4, 2)		–
C-21β		1.41 (td, 14, 5)		–
C-22α	5.57 (br s)	2.08 (ddd, 14, 4)	2.99 (d, 14)	
C-22β		1.05 (ddd, 14, 5, 2)	1.95 (d, 14)	4.82 (s)

C-23-Me	2.29 (s)	2.21 (s)	2.22 (s)	2.36 (s)
C-25-Me	1.51 (s)	1.46 (s)	1.48 (s)	1.44 (s)
C-26-Me	1.27 (s)	1.28 (s)	1.36 (s)	1.25 (s)
C-27-Me	1.44 (s)	0.56 (s)	0.89 (s)	0.98 (s)
C-28-Me	1.04 (s)	1.09 (s)	1.13 (s)	1.01 (s)
C-29-Me	—	—	—	—
C-30-Me	1.14 (d, 6.5)	—	1.36 (s)	1.11 (d, 6.5)
C-29-CH$_2$OH	—	—	—	—
C-30-CH$_2$OH	—	3.39, 3.61 (each d, 10)	—	—
C-30 OH	—	—	—	—
OCH$_3$	—	3.61 (s)	—	—

[a] Chemical shifts (δ) are expressed in parts per million from Me$_4$Si and coupling constants (J) in Hertz.
[b] Multiplicity: s = singlet, d = doublet, t = triplet, q = quartet, m = multiplet.
[c] Solvent: C$_5$D$_5$N.
[d] Solvent: CDCl$_3$.

Table 19. ¹H NMR Spectral Data of Selected Quinonemethides[a,b]

Protons on	Isoiguesterin (14) (142)	Isoiguesterol (9) (27)	Pristimerin (11) (81)	Tingenone (12) (81)
C-1	6.53 (d, 1)	6.53 (br s)	6.53 (d, 1.3)	6.55 (d, 1.2)
C-3 OH	7.01 (br s)	—[c]	6.99 (s)	7.00 (s)
C-6	7.01 (dd, 7, 1)	7.03 (br d, 7)	7.02 (dd, 7, 1.3)	7.04 (dd, 7.3, 1.2)
C-7	6.33 (d, 7)	6.37 (d, 7)	6.35 (d, 7)	6.39 (d, 7.3)
C-11α	1.97 (td, 13.5, 6)	1.84 (td, 13, 5)	1.86 (td, 13, 5)	2.03 (td, 14, 6)
C-11β	2.20 (ddd, 13.5, 5, 2)	2.12 (br d, 12)	2.16 (ddd, 13, 4.5, 2)	2.26 (ddd, 14, 4.5, 2)
C-12α	1.84 (ddd, 13.5, 6, 2)	1.66 (m)	1.80 (ddd, 13, 5, 2)	1.89 (ddd, 14, 6, 2)
C-12β	1.74 (td, 13.5, 6)	1.77 (m)	1.68 (td, 13, 4.5)	1.82 (td, 14, 4.5)
C-15α	1.74 (td, 13.5, 5)	1.74 (m)	1.66 (td, 14, 5.5)	1.815 (bd, 14, 4)
C-15β	1.52 (ddd, 13.5, 5, 2)	1.64 (m)	1.56 (ddd, 14, 6, 2)	1.66 (ddd, 14, 5, 2)
C-16α	1.43 (ddd, 13.5, 5, 2)	1.54 (m)	1.50 (ddd, 14, 5.5, 2)	1.46 (ddd, 14, 4, 2)
C-16β	1.87 (td, 13.5, 5)	1.72 (m)	1.88 (td, 14, 6)	1.92 (td, 14, 5)
C-18	1.59 (t, 4)	1.67 (m)	1.58 (d, 8)	1.67 (d, 7)
C-19α	2.41 (d, 4)	1.08 (m)	2.43 (br d, 15, 5)	2.20 (dd, 15, 7)
C-19β	—	1.80 (m)	1.69 (dd, 15.5, 8)	1.77 (ddd, 15, 13.5, 7)
C-20	—	2.10 (tq, 12, 6)	—	2.50 (ddq, 13.5, 7, 6.5)
C-21α	2.19 (ddd, 13.5, 5.2)	1.70 (m)	2.20 (ddd, 14, 4.5, 2)	—
C-21β	2.35 (br td, 13.5, 6)	1.10 (m)	1.38 (td, 14, 4)	—
C-22α	2.02 (td, 13.5, 5)	1.69 (m)	2.05 (td, 14, 4.5)	2.92 (d, 14.5)
C-22β	1.14 (br dd, 13.5, 5.5)	1.20 (m)	0.98 (ddd, 14, 4, 2)	1.86 (d, 14.5)

C-23-Me	2.21 (s)	2.22 (s)	2.21 (s)	2.33 (s)
C-25-Me	1.47 (s)	1.47 (s)	1.45 (s)	1.51 (s)
C-26-Me	1.29 (s)	1.34 (s)	1.26 (s)	1.35 (s)
C-27-Me	0.70 (s)	0.71 (s)	0.53 (s)	0.98 (s)
C-28-Me	1.16 (s)	1.19 (s)	1.10 (s)	1.02 (s)
C-30-Me	—	—	1.18 (s)	1.00 (d, 6.5)
C-29-CH_2	3.40 and 3.47 (each dd, 10, 6)	3.40 and 3.47 (each dd, 10, 6)	—	—
C-30	—	—	—	—
-CH_2OH	4.58 and 4.59 (each br s)	—	—	—
OMe	—	—	3.55 (s)	—

a Measured in $CDCl_3$. Chemical shifts (δ) are expressed in parts per million from Me_4Si and coupling constants (J) in Hertz.
b Multiplicity: s = singlet, d = doublet, t = triplet, q = quartet, m = multiplet.
c Signal not observed.

Table 20. ^1H NMR Spectral Data of Selected 14(15)-Enequinonemethide Triterpenoids[a,b]

Protons on	Balaenol (15) (140)	Balaenonol (16) (140)	Isobalaendiol (17) (140)
C-1	6.57 (d, 1)	6.57 (d, 1)	6.58 (d, 1.5)
C-3 OH	7.10 (br s)	7.10 (br s)	7.07 (br s)
C-6	7.18 (dd, 7, 1)	7.18 (dd, 7, 1)	7.19 (dd, 7, 1.5)
C-7	6.17 (d, 7)	6.21 (d, 7)	6.18 (d, 7)
C-11α	1.90 (td, 13.5, 5)	1.90 (td, 13.5, 5)	1.91 (td, 13, 5)
C-11β	1.96 (ddd, 13.5, 5, 2)	1.97 (ddd, 13.5, 5.5, 2)	1.97 (ddd, 13, 6, 2)
C-12α	1.33 (ddd, 13.5, 5, 2)	1.35 (ddd, 13.5, 5, 2)	1.31 (ddd, 13, 5, 2)
C-12β	2.48 (td, 13.5, 6)	2.40 (td, 13.5, 5.5)	2.47 (td, 13, 6)
C-16α	2.59 (d, 16.5)	2.72 (d, 16)	2.41 (br d, 16)
C-16β	1.40 (d, 16.5)	1.65 (dd, 16, 1.5)	1.44 (dd, 16, 1)
C-18	1.435 (dd, 13, 2, 5)	1.83 (ddd, 13.5, 4, 1.5)	1.71 (ddd, 12.5, 4, 1)
C-19α	1.49 (td, 13, 4)	1.99 (td, 13.5, 4)	1.00 (dt, 15, 12.5)
C-19β	1.71 (ddd)	1.87 (ddd, 13.5, 4, 3)	1.72 (m)
C-20	2.15 (m)	2.59 (qddd, 7, 6, 4, 3)	1.77 (m)
C-21	3.98 (dt, 12, 5)	4.64 (dd, 6, 1.5)	3.46 (dd, 11, 3)
C-21 OH	–	3.67 (br d, 1.5)	5.30 (s)
C-22α	1.59 (t, 12)	–	3.04 (d, 3)
C-22β	1.40 (dd, 12, 5)	–	–
C-23-Me	2.26 (s)	2.27 (s)	2.26 (s)
C-25-Me	1.27 (s)	1.29 (s)	1.28 (s)
C-26-Me	1.72 (s)	1.77 (s)	1.73 (s)
C-27-Me	0.82 (s)	0.94 (s)	0.84 (s)
C-28-Me	1.21 (s)	1.41 (s)	1.36 (s)
C-29-Me	–	–	1.03 (d, 6.5)
C-30-Me	0.95 (d, 7)	0.78 (d, 7)	–

[a] Measured in $CDCl_3$. Chemical shifts (δ) are expressed in parts per million from Me_4Si and coupling constants (J) in Hertz.
[b] Multiplicity: s = singlet, d = doublet, t = triplet, q = quartet, m = multiplet.

Partial assignment of signals in the ^{1}H-NMR spectra of several 6-oxophenolic triterpenoids have been made during their structure elucidation (41, 43). Only the aromatic, olefinic and methyl signals were assigned in this study. However, a subsequent report described complete assignment of ^{1}H-NMR spectra of zeylasteral (34) and zeylasterone (35), by application of 2D ^{1}H-^{1}H and ^{1}H-^{13}C shift-correlation techniques (139). Previous assignments made for H-1 and H-7 of these two celastroloids have been reversed and those of C-27-CH$_3$, C-28-CH$_3$ and C-30-CH$_3$ have been revised based on these studies. The unusually low chemical shifts of the carboxylic and phenolic hydroxy groups of zeylasterone (35) have been interpreted as due to strong H-bonding (Fig. 16) to the ketone

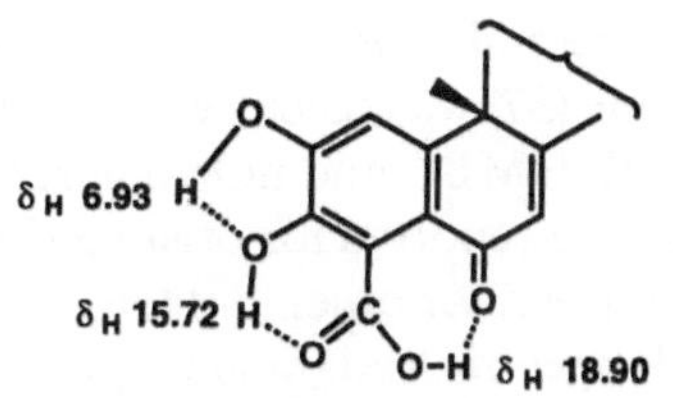

Fig. 16. Partial structure of zeylasterone (35) depicting ^{1}H-NMR chemical shifts of the H-bonded protons

and carboxylic acid carbonyl, respectively (78). The ^{1}H-NMR spectral data for zeylasteral (34) and zeylasterone (35) are presented in Table 21.

Table 21. ^1HNMR Spectral Data of Selected 6-Oxophenolic Triterpenoids[a,b]

Protons on	Zeylasteral (34) (139)	Zeylasterone (35) (139)
C-1	7.29 (s)	7.35 (s)
C-2 OH	6.36 (s)	6.93 (s)
C-3 OH	12.90 (s)	15.72 (s)
C-4-CHO	11.0 (s)	–
C-4-CO$_2$H	–	18.90 (s)
C-7	6.34 (s)	6.48 (s)
C-11α	1.93 (td, 13, 5.5)	1.95 (td, 14, 5)
C-11β	2.26 (ddd, 13, 4.5, 2)	2.35 (ddd, 14, 14.5, 2)
C-12α	1.83 (ddd, 13, 5.5, 2)	1.87 (ddd, 14, 5, 2)
C-12β	1.72 (td, 13, 4.5)	1.76 (td, 14, 4.5)
C-15α	1.70 (td, 13.5, 5.5)	1.73 (ddd, 14, 13.5, 5)
C-15β	1.58 (ddd, 13.5, 6.5, 2)	1.61 (ddd, 14, 6, 2)
C-16α	1.53 (ddd, 15, 5.5, 2)	1.56 (ddd, 15, 5, 2)
C-16β	1.91 (ddd, 15, 13.5, 6.5)	1.92 (ddd, 15, 13.5, 6)
C-18	1.61 (br d, 8)	1.63 (d, 8)
C-19α	2.43 (d, 15.5)	2.42 (br d, 15.5)
C-19β	1.70 (dd, 15.5, 8)	1.72 (dd, 15.5, 8)
C-21α	2.21 (br d, 14)	2.21 (ddd, 14, 4, 2)
C-21β	1.39 (td, 14, 4.5)	1.39 (td, 14, 4.5)
C-22α	2.05 (td, 14, 4)	2.04 (td, 14, 4)
C-22β	0.97 (br d, 14)	1.01 (ddd, 14, 4.5, 2)
C-25-Me	1.57 (s)	1.59 (s)
C-26-Me	1.32 (s)	1.34 (s)
C-27-Me	0.58 (s)	0.54 (s)
C-28-Me	1.11 (s)	1.12 (s)
C-30-Me	1.18 (s)	1.18 (s)
OMe	3.54 (s)	3.54 (s)

[a] Measured in CDCl$_3$. Chemicals shifts (δ) are expressed in parts per million from Me$_4$Si and coupling constants (J) in Hertz.
[b] Multiplicity: s = singlet, d = doublet, t = triplet.

Complete ^{1}H-NMR spectral assignment of the 7-oxoquinonemethide triterpenoid, salaciquinone (**37**) was recently made with the aid of ^{1}H-^{1}H COSY, ^{1}H-^{13}C HETCOR, HMBC and nOe difference spectroscopy and by comparison with the data reported for pristimerin (*142*). These assignments are depicted in Fig. 17. Another outcome of this study was the revision of the ^{1}H-NMR spectral assignments previously made (*112*) for the related celastroloid, dispermoquinone (**36**). ^{1}H-NMR spectra together with ^{1}H-^{1}H COSY and long-range ^{1}H-^{13}C COSY has been used to locate the hydroxy groups in ring E (*28, 120*) and ring D (*28*) of some celastroloids. Their relative stereochemical dispositions have been determined with the aid of nOe's observed (*37*) and nOe difference spectroscopy (*28*).

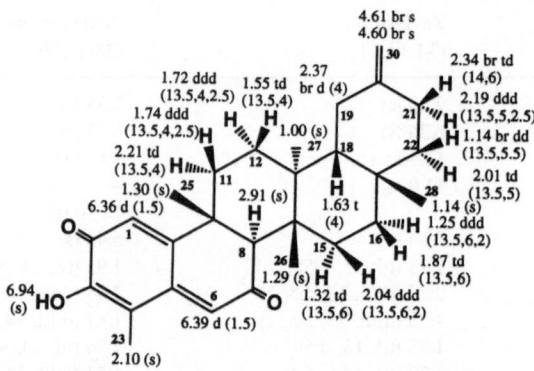

Fig. 17. ^{1}H-NMR spectral (400 MHz) assignments for salaciquinone (**37**) in CDCl$_3$

^{1}H-NMR spectroscopy using NOESY experiments was useful in determining the positions of linkages and the stereochemistry at the points of attachment of the two units in dimeric celastroloids. The ^{1}H chemical shift of the 23-CH$_3$ and the presence of nOe effects between H-1' and 23-CH$_3$ and 25-CH$_3$ in Rzedowskia bistriterpenoid-II (**43**) suggested that the position of attachment of the phenolic triterpenoid to the quinonemethide monomer is at C-4 and has α-orientation (*69*). The deshielding observed for H-6 and H-7 of the quinonemethide system further suggested that the π-orbitals of the ring A of the phenolic triterpenoid moiety resides perpendicular to the quinonemethide system. The absence of nOe effect between 23-CH$_3$ and 25-CH$_3$ in Rzedowskia bistriterpenoid-I (**42**) together with other spectroscopic data suggested that it has an epimeric relationship with Rzedowskia bistriterpenoid-II.

The nOe's observed between H-1 (or H-1′) and C-4-Me of the methylated congorosins A (**38**) and B (**41**) have suggested C(3)–O–C(2′) linkage between the two monomers in these dimeric celastroloids (*92*).

6.2.2.3.2. ¹³*C-NMR Spectroscopy.* In 1973 NAKANISHI, GOVINDACHARI and coworkers reported the application of the partially relaxed Fourier transform (PRFT) technique for the assignment of ^{13}C-NMR spectra of pristimerin (**11**), tingenin A [tingenone (**12**)], and tingenin B [22β-hydroxytingenone (**8**)] (*120*). In the case of pristimerin, assignments were made for all methyls (CH$_3$), methines (CH) and one quaternary carbon (C-20); the remaining quaternary carbons, two carbonyls and all methylenes (CH$_2$) were unassigned. It was also possible to locate the hydroxyl and carbonyl groups in tingenone and 22β-hydroxytingenone by application of PRFT ^{13}C-NMR spectroscopy. The only quaternary carbon of pristimerin assigned in the above study (*120*) was C-20 (δ 30.5), the carbon atom to which the carbomethoxy group is attached. PATRA and CHAUDHURI who assigned the ^{13}C-NMR spectra of a number of friedelanes revised the above assignment for C-20 of pristimerin and by comparison with the ^{13}C-NMR spectra of other friedelanes, they assigned the ^{13}C-NMR spectral data of pristimerin (*124*) previously reported by NAKANISHI and coworkers (*120*).

A paper published in 1989 reported the complete assignment of the ^{13}C-NMR spectra of pristimerin (**11**), tingenone (**12**) and 20α-hydroxytingenone (**6**) with the aid of ^{1}H-^{1}H and ^{1}H-^{13}C shift correlation techniques (*81*). This led to the revision of several assignments made for pristimerin by PATRA and CHAUDHURI (*124*). The broad-band proton-decoupled ^{13}C-NMR (100 MHz) spectrum of pristimerin in CDCl$_3$ which is reproduced in Fig. 18 shows signals due to all 30 carbons with the insert containing the expanded region δ 55 to 10 ppm. In a recent study, FARNSWORTH and coworkers made unambiguous ^{13}C-NMR assignments for celastrol (**1**) by a combination of ^{1}H-^{1}H DQCOSY (double quantum filtered correlation spectroscopy), APT (attached proton test), ^{1}H-^{13}C HETCOR and selective INEPT techniques (*122*). Some assignments previously made for celastrol by SNEDEN (*136*) were revised in this study.

In 1989, CORDELL, PEZZUTO and coworkers made 'unambiguous' ^{13}C-NMR spectral assignments for 22β-hydroxytingenone (**8**) by the judicious use of HETCOR, carbon satellite correlation through magnetization (CSCM 1D) and selective INEPT spectroscopy (*3*). An independent study by DIAS *et al.* (*28*) in the same year also described a complete ^{13}C-NMR assignment of 22β-hydroxytingenone, carried out as a prerequisite for the use of NMR spectral data for the structure elucidation of

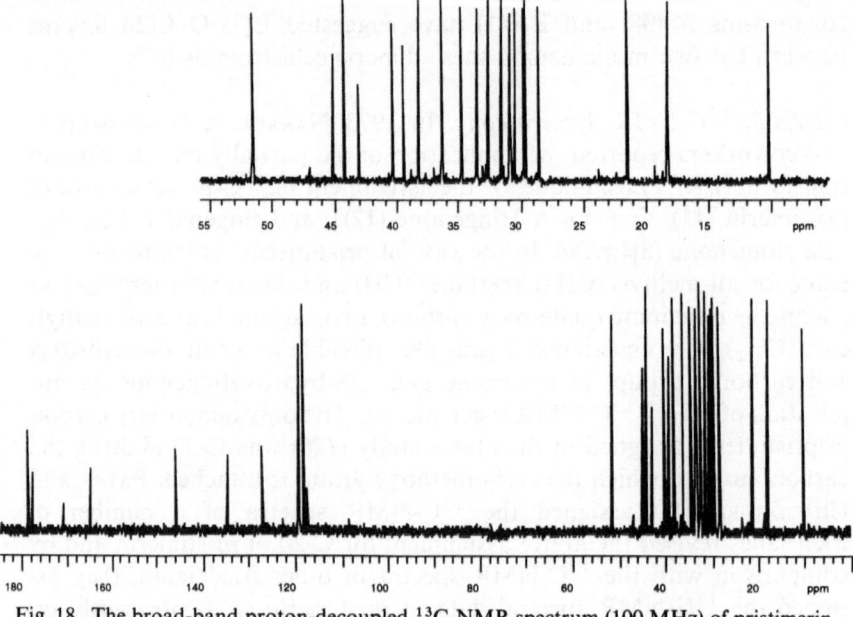

Fig. 18. The broad-band proton-decoupled ¹³C-NMR spectrum (100 MHz) of pristimerin (11) in CDCl₃

15α, 22β-dihydroxytingenone (2) (28). The assignments were made by the application of 2D ¹H-¹H, ¹H-¹³C and long range ¹H-¹³C shift correlation techniques but their assignments for C-3, C-4, C-8, C-10, C-27 and C-28 differed. Subsequent use of COSY, ROESY, HMQC and HMBC techniques by CORDELL's group suggested that their earlier assignments of C-3, C-4, C-8, C-10, C-27 and C-28 should be revised (107), their revised data agreeing well with those assignments made by the other group (28).

Detailed ¹³C-NMR analysis with the aid of 2D techniques and comparison with the revised data reported for pristimerin (81) has aided complete assignment of the ¹³C-NMR spectrum of isoiguesterin (14) (142). It also led to the revision of some assignments previously made for this quinonemethide triterpenoid (136). The ¹³C-NMR spectral assignments of excelsine (3) (11) and 15α, 22β-dihydroxytingenone (2) (28) have been made by comparing the data with those of pristimerin and 22β-hydroxytingenone, respectively. The ¹³C-NMR spectral data of some selected quinonemethides are presented in Tables 22 and 23.

Table 22. ^{13}C NMR Spectral Data of Selected Quinonemethides[a]

Carbon	Celastrol[b] (1) (122)	15α,22β-Dihydroxy-tingenone[c] (2) (28)	Excelsine[b] (3) (11)	30-Hydroxy-pristimerin[b] (5) (27)	20α-Hydroxy-tingenone[b] (6) (81)	20-Hydroxy-20-epi-tingenone[b] (7) (122)
C-1	120.59 d	121.0 d	120.2 d	119.6 d	119.8 d	119.59 d
C-2	178.20 s	179.2 s	178.2 s	178.4 s	178.4 s	178.30 s
C-3	146.99 s	148.2 s	146.0 s	146.0 s	146.2 s	146.00 s
C-4	120.64 s	118.1 s	117.8 s	117.1 s	117.1 s	117.17 s
C-5	127.40 s	128.1 s	127.4 s	127.5 s	127.9 s	127.56 s
C-6	135.62 d	132.6 d	135.2 d	133.9 d	133.3 d	133.64 d
C-7	118.26 d	120.6 d	118.2 d	118.2 d	118.3 d	118.26 d
C-8	172.95 s	163.9 s	169.7 s	169.8 s	168.7 s	168.92 s
C-9	43.04 s	42.6 s	43.9 s	42.9 s	42.9 s	42.84 s
C-10	164.99 s	164.2 s	165.0 s	164.7 s	164.2 s	164.22 s
C-11	33.72 t	33.5 t	33.9 t	33.6 t	33.2 t	33.02 t
C-12	29.21 t	31.4 t	29.3 t	29.8 t	29.9 t	29.68t
C-13	39.22 s	39.6 s	39.4 s	39.4 s	40.0 s	39.82 s
C-14	45.27 s	48.9 s	45.3 s	45.0 s	44.2 s	44.04 s
C-15	28.64 t	72.3 d	29.5 t	28.6 t	29.4 t	29.15 t
C-16	36.27 t	37.4 t	36.3 t	36.4 t	35.7 t	35.50 t
C-17	30.59 s	44.5 s	32.3 s	30.8 s	35.9 s	35.91 s
C-18	44.16 d	45.6 d	47.9 d	43.6 d	43.3 d	43.15 d

Table 22 (continued)

Carbon	Celastrol[b] (1) (122)	15α,22β-Dihydroxy-tingenone[c] (2) (28)	Excelsine[b] (3) (11)	30-Hydroxy-pristimerin[b] (5) (27)	20α-Hydroxy-tingenone[b] (6) (81)	20-Hydroxy-20-epi-tingenone[b] (7) (122)
C-19	30.99 t	32.2 t	36.3 t	25.5 t	36.9 t	36.72 s
C-20	39.86 s	41.2 d	38.5 s	46.2 s	73.7 s	73.62 s
C-21	29.39 t	214.0 s	79.0 d	25.1 t	214.9 s	215.06 s
C-22	34.40 t	78.8 d	43.5 t	34.0 t	50.5 t	50.31 t
C-23	10.45 q	10.5 q	10.4 q	10.3 q	10.3 q	10.25 q
C-25	38.32 q	41.0 q	38.3 q	38.4 q	38.5 q	38.40 q
C-26	21.37 q	23.9 q	21.5 q	21.7 q	23.3 q	23.04 q
C-27	18.61 q	24.4 q	18.8 q	18.4 q	19.4 q	19.33 q
C-28	31.40 q	25.8 q	31.9 q	31.5 q	33.2 q	33.02 q
C-29	182.65 s	—	73.0 t	177.4 s	—	—
C-30	32.33 q	15.2 q	24.5 q	74.1 t	29.0 q	28.86 q
OCH$_3$	—	—	—	51.9 q	—	—

[a] Chemical shifts are reported in δ units from internal standard Me$_4$Si.
[b] Solvent: CDCl$_3$.
[c] Solvent: C$_5$D$_5$N.

Table 23. ^{13}C NMR Spectral Data of Selected Quinonemethides[a]

Carbon	22β-Hydroxy-tingenone[b] (8) (28)	Isoiguesterol[c] (9) (27)	Pristimerin[c] (11) (81)	Tingenone[c] (12) (81)	Isoiguesterin[c] (14) (142)
C-1	121.4 d	119.4 d	119.6 d	119.8 d	119.6 d
C-2	179.4 s	178.4 s	178.4 s	178.4 s	178.3 s
C-3	148.3 s	146.0 s	146.1 s	146.1 s	146.0 s
C-4	117.8 s	117.2 s	117.0 s	117.1 s	117.1 s
C-5	128.2 s	127.4 s	127.5 s	127.8 s	127.4 s
C-6	132.3 d	134.2 d	133.9 d	133.5 d	133.9 d
C-7	118.3 d	118.2 d	118.1 d	118.1 d	117.9 d
C-8	166.9 s	170.7 d	169.9 s	168.6 s	170.1 s
C-9	42.3 s	43.3 s	42.9 s	42.7 s	42.9 s
C-10	163.9 s	164.4 s	164.7 s	164.7 s	165.0 s
C-11	34.3 t	33.1 t	33.6 t	33.8 t	33.9 t
C-12	29.8 t	29.4 t	29.7 t	30.0 t	29.7 t
C-13	40.7 s	40.0 s	39.4 s	40.6 s	41.3 s
C-14	44.3 s	44.0 s	45.0 s	44.7 s	44.8 s
C-15	28.5 t	29.7 t	28.7 t	28.5 t	.28.4 t
C-16	29.7 t	36.5 t	36.4 t	35.5 t	36.0 t
C-17	44.9 s	30.2 s	30.6 s	38.2 s	31.6 s
C-18	45.2 d	43.7 d	44.4 d	43.6 d	44.9 d
C-19	32.1 t	25.2 t	30.9 t	32.1 t	30.4 t
C-20	41.2 d	33.0 d	40.4 s	41.9 d	147.9 s
C-21	213.4 s	22.5 t	29.9 t	213.5 s	30.5 t
C-22	77.1 d	35.2 t	34.8 t	52.5 t	36.9 t
C-23	10.5 q	10.3 q	10.2 q	10.3 q	10.2 q
C-25	39.0 q	37.7 q	38.3 q	39.1 q	38.9 q
C-26	21.7 q	23.4 q	21.6 q	21.6 q	21.3 q
C-27	20.3 q	17.9 q	18.3 q	19.7 q	19.7 q
C-28	25.5 q	36.2 q	31.6 q	32.6 q	31.1 q
C-29	–	69.3 q	178.7 s	–	108.2 t
C-30	15.1 q	–	32.7 q	15.1 q	–
OCH$_3$	–	–	51.5 q	–	–

[a] Chemical shifts are reported in δ units from internal standard Me$_4$Si.
[b] Solvent: C$_5$D$_5$N.
[c] Solvent: CDCl$_3$.

^{13}C-NMR spectral assignments of the netzahualcoyone series of 14(15)-enequinonemethide triterpenoids have been reported (65). Although the basis of these assignments was not discussed, the assignments made correlate well with those obtained for the balaenol series though the application of HETCOR and long-range HETCOR techniques (140). The ^{13}C-NMR spectral assignments for some selected 14(15)-enequinonemethides are listed in Table 24.

Table 24. ¹³C NMR Spectral Data of Selected 14(15)-Enequinonemethide Triterpenoids[a]

Carbon	Balaenol (15) (140)	Balaenonol (16) (140)	Isobalaendiol (17) (140)	Netzahualcoyene[b] (19) (65)	Netzahualcoyol[b] (20) (65)	Netzahualcoyondiol[b] (21) (65)
C-1	120.0 d	120.2 d	120.0 d	121.69	121.59	121.68
C-2	178.1 s	178.2 s	178.1 s	178.14	178.16	178.32
C-3	146.3 s	146.4 s	146.3 s	116.94	117.27	116.91
C-4	116.8 s	116.7 s	116.9 s	128.43	127.71	128.12
C-5	128.2 s	128.2 s	127.8 s	146.38	146.40	146.47
C-6	134.7 d	134.1 d	134.5 d	134.96	135.34	134.63
C-7	121.6 d	122.2 d	121.6 d	120.02	120.04	120.25
C-8	160.1[c] s	159.6[c] s	160.1[c] s	158.80	159.65	159.05
C-9	44.5 s	44.2 s	44.6 s	42.73	41.16	42.68
C-10	159.6[c] s	157.6[c] s	159.4[c] s	160.19	160.27	160.16
C-11	37.5 t	37.3 t	37.5 t	34.05	35.46	34.10
C-12	36.0 t	36.1 t	35.2 t	35.74	35.86	35.06
C-13	42.7 s	42.1 s	42.5 s	44.60	44.83	44.05
C-14	135.8 s	136.4 s	136.4 s	135.44	134.81	136.05
C-15	127.7 s	126.5 s	126.8 s	127.68	111.98	126.83
C-16	39.4 t	38.4 t	37.7 t	37.64	37.16	37.67
C-17	_[d]	49.4 s	40.0 s	43.17	54.13	47.63
C-18	42.3 d	46.1 d	43.1 d	44.05	48.54	39.09
C-19	32.2 t	30.6 t	34.3 t	29.74	39.19	38.16
C-20	34.5 d	35.9 d	33.0 d	33.88	49.31	44.73

C-21	68.3 d	73.91 d	74.2 d	36.20	77.36	79.65
C-22	43.4 t	215.3 s	80.0 d	37.91	29.79	69.50
C-23	10.4 q	10.4 q	10.4 q	10.46	10.53	10.55
C-25	29.3 q	28.97 q	29.4 q	29.53	29.17	29.66
C-26	22.0 q	22.0 q	21.9 q	24.08	24.35	24.78
C-27	24.3 q	24.3 q	24.4 q	22.04	22.60	21.99
C-28	31.6 q	22.3 q	27.4 q	19.88	174.21^c	14.22
C-29	—	—	—	179.37	175.27^c	179.25
C-30	10.6 q	11.5 q	18.6 q	31.57	22.01	27.25
OCH$_3$	—	—	—	51.88	52.33	52.44

a Measured in CDCl$_3$. Chemical shifts reported in δ units from internal standard Me$_4$Si.
b Multiplicity not recorded.
c Chemical shifts in the same column are interchangeable.
d Not assigned due to minor impurity signals.

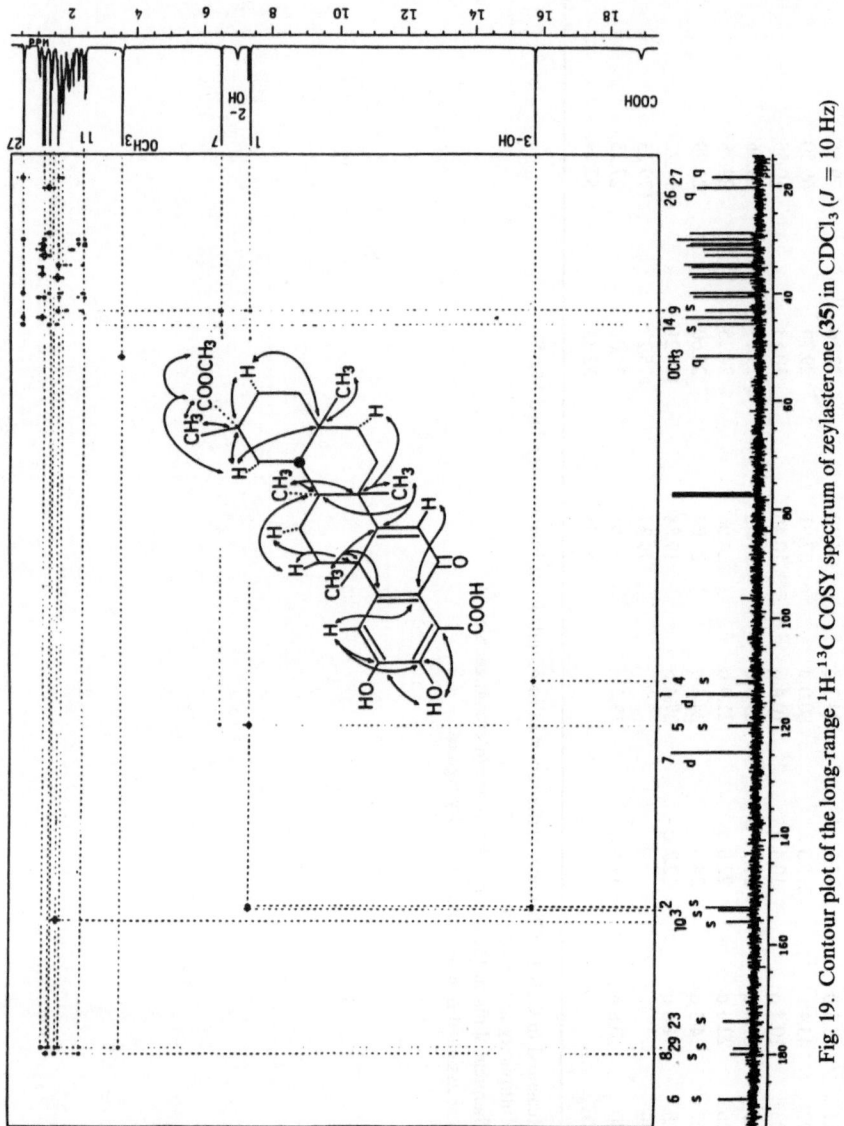

Fig. 19. Contour plot of the long-range ^{1}H-^{13}C COSY spectrum of zeylasterone (35) in CDCl$_3$ ($J = 10$ Hz)

Table 25. ^{13}C NMR Spectral Data of Selected Phenolic Triterpenoids[a]

Carbon	Isopristrimerin-III[b] (27) (93)	Isotingenone-III[b] (28) (93)	Zeylasteral (34) (139)	Zeylasterone (35) (139)
C-1	108.75	109.93	116.2 d	113.8 d
C-2	143.48	144.40	149.2 s	152.9 s
C-3	142.01	144.28	149.7 s	153.5 s
C-4	120.27	121.25	117.0 s	111.4 s
C-5	124.48	124.40	122.9 s	119.5 s
C-6	122.42	122.01	186.0 s	188.1 s
C-7	138.61	137.49	125.3 d	124.4 d
C-8	44.16	44.40	173.8 s	179.8 s
C-9	143.78	146.30	40.52[c] s	42.9 s
C-10	130.32	130.08	150.5 s	155.6 s
C-11	119.80	120.85	33.8 t	34.4 t
C-12	33.35	32.54	29.81[d] t	29.84 t
C-13	40.87	40.35	39.4 s	39.7 s
C-14	41.07	41.02	45.1 s	45.6 s
C-15	24.18	24.35	28.7 s	28.6 t
C-16	37.36	36.23	36.4 t	36.2 t
C-17	31.87	39.21	30.6 s	30.6 s
C-18	47.37	42.41	44.4 d	44.3 d
C-19	31.77	37.69	30.9 t	30.9 t
C-20	39.30	45.90	40.49[c] s	40.5 s
C-21	30.10	213.23	29.85[d] t	29.78 t
C-22	38.03	51.42	34.9 t	34.9 t
C-23	11.08	11.95	200.3 d	173.8 s
C-25	22.63	22.77	36.4 q	36.8 q
C-26	19.55	20.64	20.5 q	20.2 q
C-27	19.10	19.82	18.3 q	18.3 q
C-28	33.03	31.45	31.7 q	31.6 q
C-29	180.24	–	178.8 s	178.7 s
C-30	30.83	15.72	32.8 q	32.7 q
CO_2CH_3	51.91	–	51.5 q	51.6 q

[a] Measured in $CDCl_3$. Chemical shifts are reported in δ units from internal standard Me_4Si.
[b] Multiplicity not recorded.
[c,d] Chemical shifts in the same column are interchangeable.

The ^{13}C-NMR spectrum of zeylasterone (35), the first 6-oxophenolic triterpenoid to be encountered in nature, was assigned by comparison with ^{13}C-NMR data reported for friedelanes and some aromatic and O-heterocyclic compounds (78). As more ^{13}C-NMR data of friedelanes became available, comparisons were made which suggested that the previous assignments for C-2 and C-8, and C-27 and C-28 should be

Table 26. $^{13}C\,NMR$ Spectral Data of Selected Dimeric Celastroloids[a]

Carbon	Rzedowskia bistriterpenoid-I (42) (69)		Rzedowskia bistriterpenoid-II (43) (69)		Umbellatin α (44) (72)	
	C	C'	C	C'	C	C'
C-1	110.8 d	114.9 d	110.3 d	115.3 d	110.49	115.05
C-2	179.2 d	173.4 s	188.0 s	174.0 s	188.40	174.00
C-3	171.3 s	145.3 s	171.5 s	144.7 s	172.66	170.05
C-4	91.2 s	124.0 s	92.1 s	124.0 s	91.16	123.89
C-5	128.5 s	132.0 s	127.7 s	130.1 s	128.59	132.24
C-6	129.0 d	189.6 s	126.7 d	189.0 s	126.17	189.40
C-7	117.4 d	126.3 d	116.2 d	126.2 d	115.05	128.46
C-8	164.5 s	151.3 s	161.4 s	150.5 s	163.50	151.15
C-9	38.8 s	44.0 s	38.2 s	41.9 s	38.65	43.63
C-10	137.7 s	151.3 s	137.7 s	151.0 s	137.00	151.17
C-11	33.0 t	34.1 t	33.0 t	34.2 t	33.16	33.16
C-12	29.5 t	29.7 t	29.7 t	29.8 t	29.65	29.90
C-13	39.1 s	39.3 s	39.0 s	39.9 s	41.85	39.83
C-14	44.5 s	44.5 s	44.7 s	44.2 s	44.30	44.01
C-15	28.7 t	29.4 t	28.7 t	28.6 t	29.85	28.52
C-16	36.5 t	36.5 t	36.4 t	36.5 t	35.53	34.17
C-17	30.7 s	30.7 s	30.6 s	30.6 s	38.10	38.16

C-18	44.8 d	44.8 d	44.5 d	44.7 d	43.60	43.60
C-19	30.9 t	31.0 t	30.9 t	31.0 t	30.14	31.92
C-20	40.6 s	40.7 s	40.5 s	40.5 s	41.85	41.85
C-21	29.8 t	30.0 t	29.9 t	29.9 t	—[b]	—[b]
C-22	34.9 t	35.1 t	34.9 t	34.9 t	52.39	52.42
C-23	24.7 q	13.3 q	22.5 q	12.8 q	13.11	23.85
C-25	37.8 q	40.2 q	37.6 q	37.7 q	40.20	39.82
C-26	21.0 q	22.5 q	20.9 q	22.2 q	20.82	22.30
C-27	18.3 q	18.6 q	18.3 q	18.6 q	19.59	19.76
C-28	31.7 q	31.7 q	31.6 q	31.6 q	32.58	32.58
C-29	179.0 s	179.0 s	179.0 s	179.0 s	—	—
C-30	33.0 q	33.0 q	32.7 q	32.8 q	15.06	15.06
OCH_3	51.6 q	51.8 q	51.5 q	51.5 q	—	

[a] Measured in $CDCl_3$. Chemical shifts are reported in δ units from internal standard Me_4Si.
[b] Signal not recorded.

revised (*124*). Recent application of ^{1}H-^{13}C COSY and ^{1}H-^{13}C long range correlation NMR spectroscopy aided more reliable assignment of the ^{13}C-NMR spectra of zeylasterone (**35**) and the related compound, zeylasteral (**34**) (*139*). The contour plot of the ^{1}H-^{13}C long-range correlation spectrum of zeylasterone (**35**) is reproduced in Fig. 19 and was especially useful in the unambiguous assignment of quaternary carbons other than C-6 (which was assigned by chemical shift arguments). The correlations observed are diagrammatically represented in Fig. 19. ^{13}C-NMR spectral assignments of some selected phenolic triterpenoids, including 6-oxophenolic triterpenoids are given in Table 25. The 2D NMR (HETCOR and HMBC) experiments along with the chemical shift arguments and signal multiplicities have been used in the assignments of the ^{13}C-NMR spectrum of the 7-oxoquinonemethide triterpenoid, salaciquinone (**37**) (*142*). These assignments are depicted in Fig. 20. 2D NMR analysis, especially HMBC, was helpful in ruling out the possible isomeric structures for celastranhydride (**46**) (*46*).

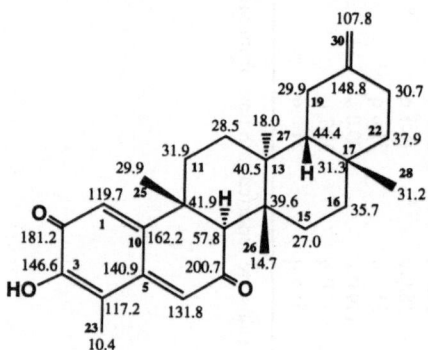

Fig. 20. ^{13}C-NMR spectral (100 MHz) assignments for salaciquinone (**37**) in CDCl$_3$

^{13}C-NMR spectroscopy has played a useful role in the structure elucidation of dimeric celastroloids (*69, 72, 92*). Thus, the use of ^{1}H-^{13}C long-range coupling system in COLOC spectrum established the linkage between the two monomeric units in cangorosin B (**41**). The ^{13}C-NMR data of Rzedowskia bistriterpenoids-I (**42**), -II (**43**), and umbellatin α (**44**) are presented in Table 26.

6.2.2.4. Mass Spectrometry

Ironically, mass spectrometric studies had not played an important role in the early structural investigations of celastrol and pristimerin.

However, NAKANISHI and coworkers utilized MS data to unambiguously support the structures of the diacetates derived from the acid-catalyzed isomerization products of pristimerin, namely, isopristimerin-I and -II (*88*). In a subsequent paper, NAKANISHI, TAKAHASHI and BUDZIKIEWICZ published the actual mass spectra and postulated fragmentation pathways of these and several other derivatives of pristimerin (*119*). These postulates formed the basis of fragmentation pathways proposed later for other celastroloids and their derivatives.

During the structure elucidation of tingenone (**12**), THOMSON, SESHADRI and coworkers observed that its MS was virtually identical with that of pristimerin at m/z values below 300 (*9*). The presence of significant peaks at m/z 267, 253, 241, 227, 202, 201, and 200 suggested that the major fragmentations occurred at the C/D ring junction with hydrogen transfers occurring analogous to friedelane derivatives (*17*). The electron-impact MS (EI-MS) of pristimerin is reproduced in Fig. 21, and the fragmentation pathways leading to some significant ions in quinonemethide triterpenoids are depicted in Scheme 1.

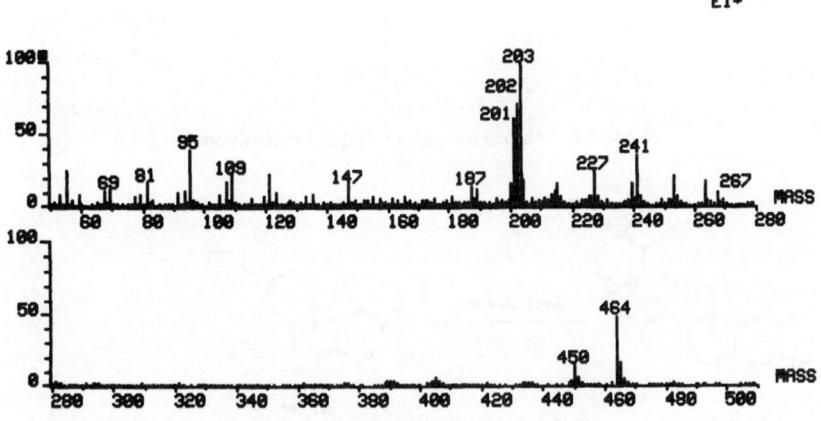

Fig. 21. The EI-MS of pristimerin (**11**)

The MS of the 14(15)-enequinonemethides, balaenol (**15**), balaenonol (**16**), and isobalaendiol (**17**) showed significant peaks at m/z 294, 279, and 253 arising from a *retro*-Diels-Alder cleavage of ring D followed by the loss of a CH_3 or a C_3H_5 group (Scheme 2) (*37*). The MS peak at m/z 279 appears to be common to the 14(15)-enequinonemethide, netzahualcoyene (**19**) (*65*) and the 9(11)-enequinonemethides, pristimerinene (**25**) and hydroxypristimerinene (**24**) (*24*) (see Sect. 10). 23-Oxoisopristimer-

in-III (29), one of the few known naturally occurring phenolic triterpenoids, has been reported to undergo a typical *retro*-Diels-Alder fragmentation of ring C. As illustrated in Scheme 3, the resulting two halves
underwent further fragmentations to produce significant ions which may
be useful in the structure elucidation of this class of celastroloids (*41*). The

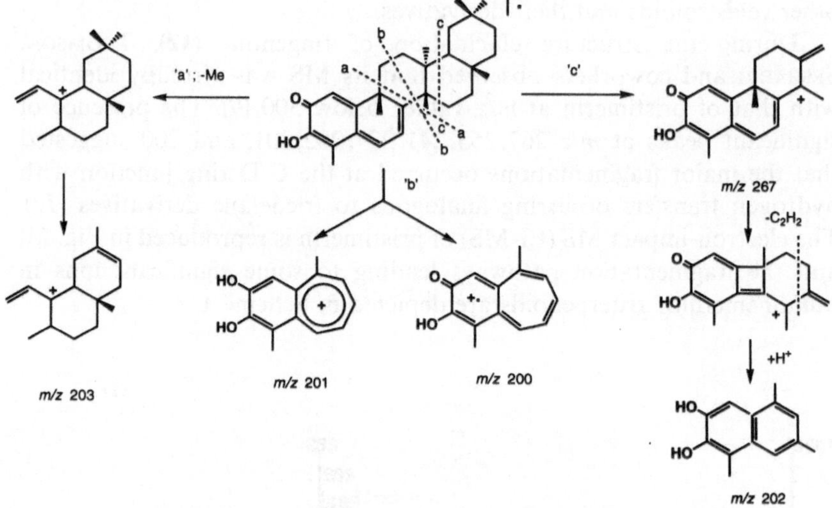

Scheme 1. MS fragmentation of quinonemethides

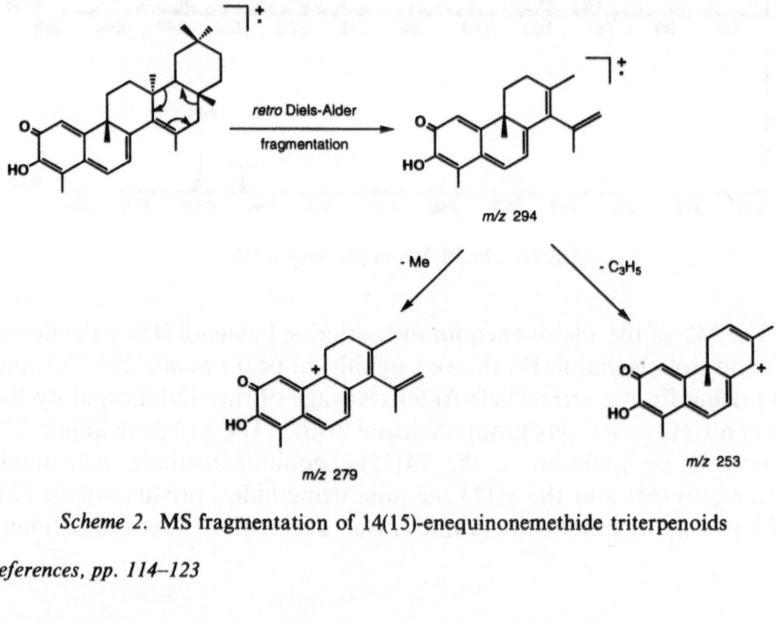

Scheme 2. MS fragmentation of 14(15)-enequinonemethide triterpenoids

presence of a base peak at m/z 229 in the MS of the 6-oxophenolic triterpenoid, zeylasterone (**35**) has been explained by invoking the fragmentation pathway depicted in Scheme 4 (*78*).

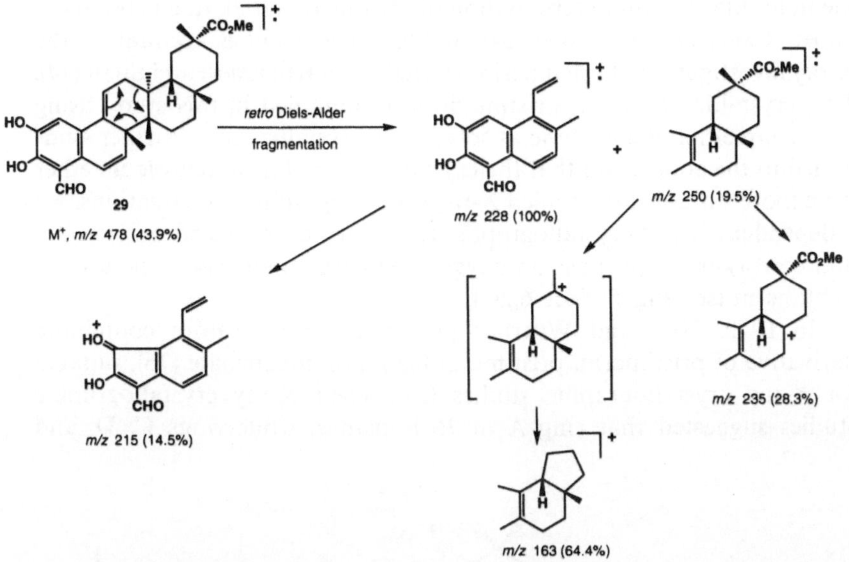

Scheme 3. MS fragmentation of the phenolic triterpenoid, 23-oxoisopristimerin-III (**29**)

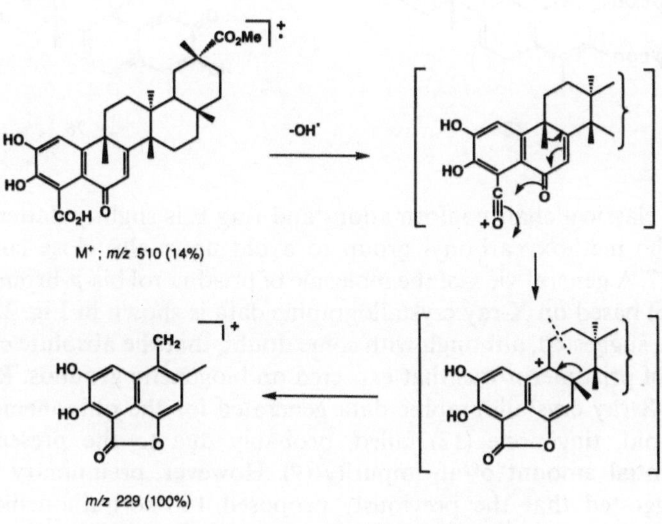

Scheme 4. MS fragmentation of zeylasterone (**35**)

6.2.2.5. X-Ray Crystallography

Early attempts to apply X-ray crystallography to structure elucidation of celastroloids met with mixed results. A communication by KULKARNI and SHAH in 1954 described the application of X-ray method for the molecular weight determination of pristimerin (*102*). In a subsequent paper, CARLISLE and EHRENBERG published a detailed account of the X-ray investigation of pristimerin referred to by KULKARNI and SHAH (*14*). Two crystalline forms of pristimerin were obtained in this study using petroleum ether and acetone as solvents of crystallization. Further studies led to the conclusion that the crystals obtained from petroleum ether were more suitable for detailed X-ray crystallographic investigations. An independent X-ray crystallographic study, however, misled BHATNAGAR and coworkers to propose an oxygenated carotenoid type structure for pristimerin (see Fig. 5; Sect. 6.2.1).

In 1972, HAM and WHITING prepared a heavy atom containing derivative of pristimerin, pristimerol bis-*p*-bromobenzoate (**75**), suitable for X-ray crystallographic studies (*87*). Their X-ray crystallographic studies suggested that ring A in **75** is planar, while rings C, D, and

75   **76**

E adopt classical chair conformations and ring E is slightly flattened to enable the methoxycarbonyl group to avoid unusually close contacts with C-27. A general view of the molecule of pristimerol bis-*p*-bromobenzoate (**75**) based on X-ray crystallographic data is shown in Fig. 22. The data also suggested, although with some doubt, that the absolute configuration of pristimerin was that expected on biogenetic grounds. Refinement of X-ray crystallographic data generated for the quinonemethide triterpenoid, tingenone (**12**) failed, probably due to the presence of a substantial amount of an impurity (*9*). However, preliminary X-ray data suggested that the previously proposed 16-oxo-quinonemethide structure (**76**) for tingenone (*100*) was incorrect.

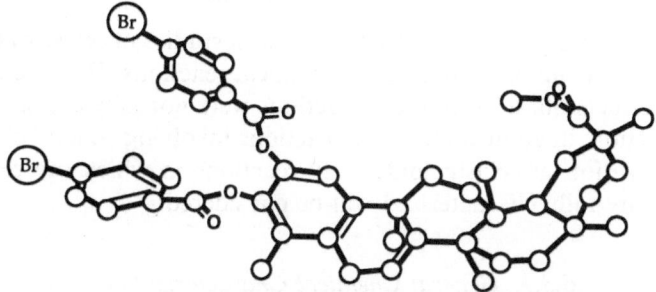

Fig. 22. A general view of pristimerol bis-*p*-bromobenzoate (**75**) molecule based on its X-ray crystallographic structure

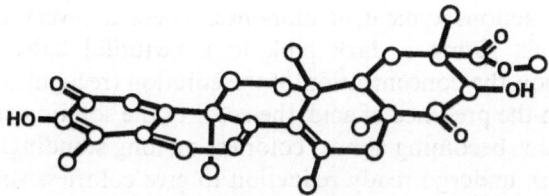

Fig. 23. A general view of the molecule of netzahualcoyone (**22**) based on its X-ray crystallographic structure

In the early 1980's GONZALES and coworkers determined the molecular structure and absolute configuration of netzahualcoyone (**22**) by X-ray crystallography (*61*). The study was undertaken to differentiate between the two possible locations for the additional double bond [9(11) or 14(15)] of this enequinonemethide triterpenoid. The X-ray data suggested that netzahualcoyone had the new carbon skeleton of a rearranged pristimerin with 14(15)-unsaturation and a *cis* D/E ring fusion. Rings A and B were found to be planar, while rings C and E had chair conformation and ring D had an envelope type conformation. A comparative conformational study between netzahualcoyone (**22**) and pristimerol bis-*p*-bromobenzoate (**75**) has also been carried out by these workers. It was found that the presence of the double bond at C-14(15) in **22** instead of the methyl group at C-14 in **75** caused the torsion C(16)-C(17)-C(18)-C(13) to change from 57° in netzahualcoyone (**22**) to − 43° is pristimerol bis-*p*-bromobenzoate (**75**). The conformational differences observed for rings C, D, and E of these two compounds were attributed to the change in this torsion angle. A general view of the molecule of netzahualcoyone based on its X-ray crystallographic data is presented in Fig. 23.

6.3. Chemical Reactions

As noted in Sect. 6.2.1, the early structural studies of celastrol (**1**) and pristimerin (**11**) involved numerous chemical reactions. However, many of the products obtained in these reactions were not fully characterized. In this section, in addition to some reactions involving general chemical characterization of celastroloids, only reactions in which the resulting products are fully characterized will be considered.

6.3.1. General Chemical Characterization

Celastroloids in general show the typical color reaction of phenols, *e.g.* formation of green coloration with neutral iron(III) chloride (*74, 98, 135*). The quinonemethide and enequinonemethide classes, in addition, show color reactions typical of quinones. These are very sensitive to alcoholic alkali, giving a dark pink to a beautiful light pink color depending upon the concentration of the solution (recognizable even in traces) (*98*). In the presence of acid, the color of the solution slowly fades to a light yellow, becoming almost colorless on long standing. Like other quinones, these undergo ready reduction to give colorless solutions but gradually gain color on exposure to air. Reducing agents which have been used for this purpose include hydrogen/catalyst (*48, 118, 134*), sulfurous acid (*48*), and sodium borohydride (*118*).

Pristimerin (**11**) does not combine with sodium bisulfite in an aqueous medium, but does so easily in the presence of alcohol, giving a colorless addition compound (*40, 98*). With acetic anhydride, pristimerin gives an orange color, which on addition of boroacetic anhydride (Dimroth reagent), turns pink; the color fades on boiling (*40, 98*). Treatment of pristimerin with cold acetic anhydride containing a trace of sulfuric acid (Thiele reagent), gives a deeper shade of red, but the color gradually fades resulting in a colorless solution (*98*) (see Sect. 6.3.5).

The formation of an ether soluble deep green copper chelate upon treatment of an ethereal solution of pristimerin with saturated aqueous copper acetate has been reported (*73*). Pristimerin also forms chelates with nickel and aluminium. Although the structures of these chelates have not been determined, their formation suggested the presence of strong hydrogen bonding between a carbonyl and a hydroxyl group in the parent quinonemethide.

6.3.2. Degradation and Oxidation

A wide variety of degradation and oxidation reactions were used during the structure elucidation of celastrol and pristimerin. Only those studies which have provided useful structural information will be discussed in this section. Controlled degradation of pristimerin was useful in establishing the nature of its carbocyclic skeleton. Thus, KMnO₄ oxidation of pristimerin (11) afforded an ester anhydride (77) which upon selenium dehydrogenation gave 1,2,6-trimethylphenanthrene (78) (Scheme 5). This evidence coupled with the production of alkylpicene by zinc-dust distillation has aided in establishing the presence of a pentacyclic triterpene ring system in pristimerin (*118*). The structure of the ester anhydride (77) was elucidated by ¹H-NMR spectroscopy (*95*).

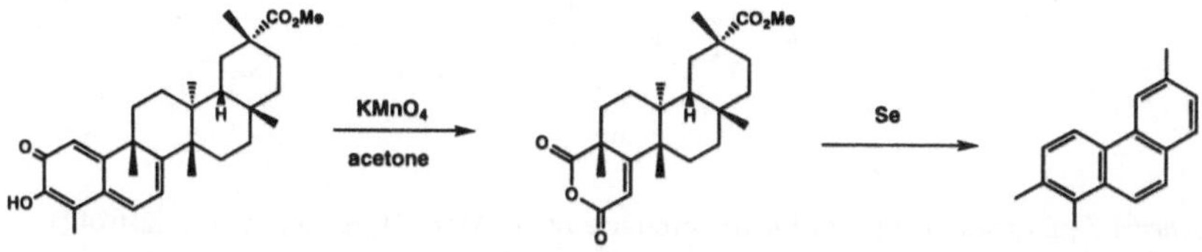

Scheme 5. Oxidative degradation of pristimerin (11)

GISVOLD found that celastrol underwent ozonolysis producing a keto acid with loss of three carbon atoms (*50*) which yielded a 2,4-DNP derivative but was not adequately characterized. Ozonolysis of pristimerin was attempted by NAKANISHI and his coworkers who failed to detect the formation of any volatile product (*118*). SESHADRI et al. reported the formation of a crude ozonide, which on decomposition with water, afforded a steam volatile substance. Based on the molecular formula of the 2,4-DNP derivative, the steam volatile substance was suspected to be glycollic aldehyde (*134*). Ozonolysis has also been used to establish the presence of a $\rangle C(Me)CO_2Me$ moiety in pristimerin. Thus, dimethylpristimerol (55; Table 11) (obtained by LiAlH₄ reduction of pristimerin followed by methylation) upon treatment with PCl₅ followed by ozonolysis produced acetaldehyde in 15% yield (*95*). JOHNSON and his collaborators argued that such a result can be rationalized (on the basis of a triterpene skeleton) only if pristimerol contains partial structure 79, which by reduction and rearrangement would give 80 together with its double bond isomers (see Scheme 6). Evidence for the presence of a $\rangle C(Me)CO_2H$ moiety in celastrol (1) was provided by a different sequence of reactions which involved treating celastrol reductive acetate (81) with Pb(OAc)₄ affording a decarboxylated *nor*-analog (see Scheme

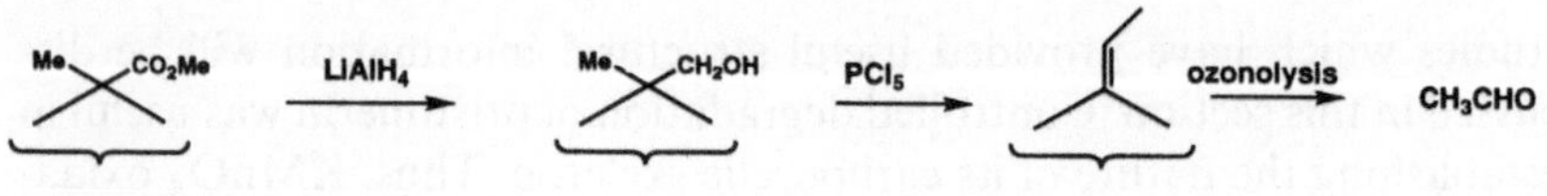

Scheme 6. Degradative proof for the presence of a $>$C(Me)CO_2Me moiety in pristimerin
(11)

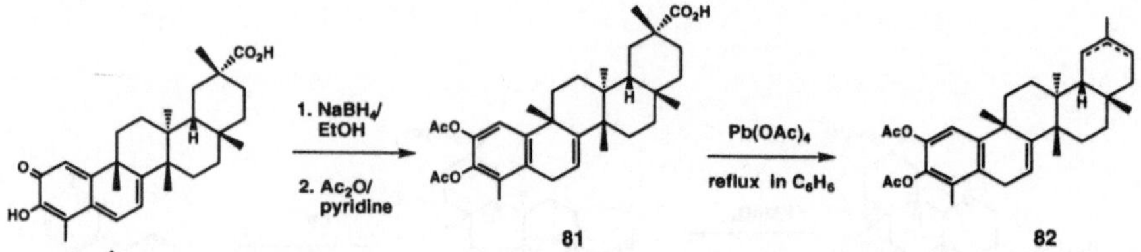

Scheme 7. Degradative proof for the presence of $>$C(Me)CO_2H moiety in celastrol (1)

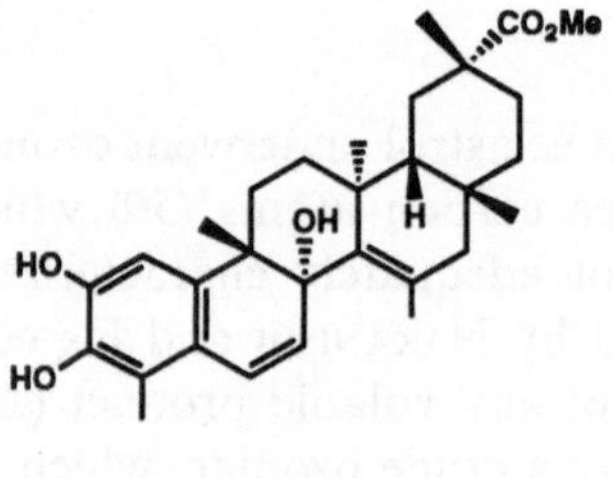

7), the structure of which was elucidated as **82** with the aid of its
¹H-NMR spectrum (119).

DDQ (2,3-dichloro-5,6-dicyanobenzoquinone) oxidation of pristimerin (**11**) has afforded four products, the 14(15)-enequinonemethide,
netzahualcoyene (**19**), the alcohol **83** and two dimeric products identical
with Rzedowskia bistriterpenoids-I (**42**) and -II (**43**) (69). Analogous
treatment of tingenone (**12**) with DDQ has resulted in, albeit a poor yield,
in the two naturally occurring dimeric quinonemethide-6-oxophenolic
celastroloids, umbellatin α (**44**) and umbellatin β (**45**) (72). The formation
of these dimers from quinonemethide precursors suggests the inter-
mediacy of a 6-oxophenolic triterpenoid as depicted in Scheme 8.

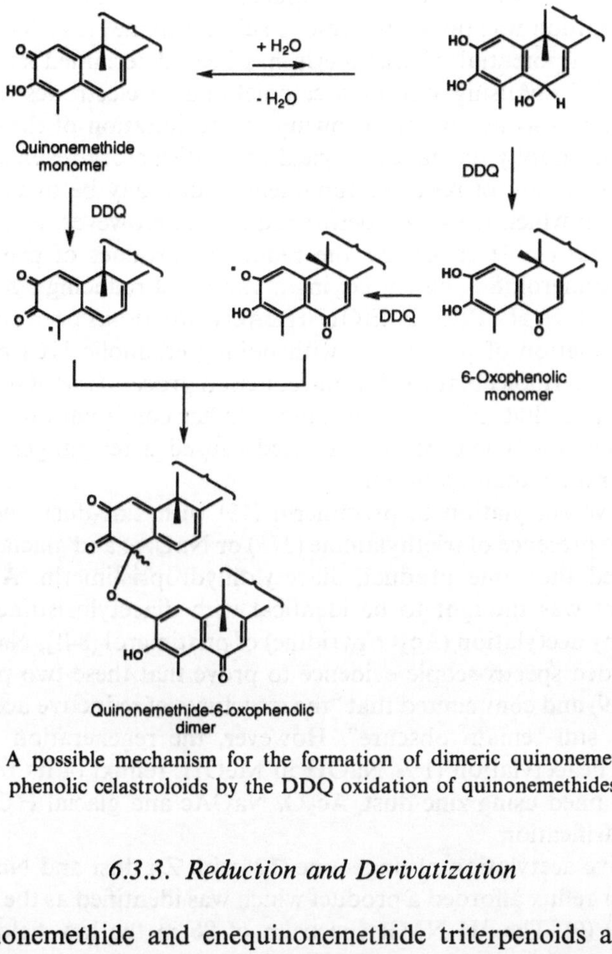

Scheme 8. A possible mechanism for the formation of dimeric quinonemethide-6-oxo-phenolic celastroloids by the DDQ oxidation of quinonemethides

6.3.3. Reduction and Derivatization

Quinonemethide and enequinonemethide triterpenoids are readily reduced by a variety of reagents (see Section 6.3.1). Reduction of the quinonemethide system with reagents such as $NaBH_4$ and $LiAlH_4$ involves a simple nucleophilic addition of a hydride ion at C-6 resulting in a ring A aromatic system. Some derivatives derived from these reduction reactions have assisted in structure elucidation and semisynthesis of natural celastroloids.

The ability of celastrol to undergo reduction with hydrogen in the presence of Raney-Ni as a catalyst was first observed by GISVOLD in 1939 (47). However, attempts to isolate the reduction product proved unsuccessful as it underwent ready reoxidation back to celastrol. In 1942, SCHECHTER and HALLER observed that celastrol and pristimerin were readily reduced by both catalytic hydrogen and sulfurous acid and that the original color was restored on exposure to air or in the latter case when the solution was boiled to remove sulfur dioxide (*132*). The oxidation-reduction potential of pristimerin has been determined as approximately $+297$ mv using saturated calomel and Pt electrodes (*118*). An accurate value was not obtained owing to precipitation of the quinhydrone. Polarography has failed to yield any satisfactory results.

Aerial oxidation of reduced quinonemethides may be overcome by derivatization which is usually performed *in situ*. However, Nakanishi's group has been able to isolate the reduction product of pristimerin, namely pristimerol (**84**), by employing a variety of reducing agents (e.g. NaBH$_4$ in EtOH, H$_2$/PtO$_2$ in EtOH-HOAc) (*118*). It has been suggested that the formation of pristimerol with boiling ethanolic HCl proceeds through an intermolecular oxidation-reduction process (*118*). Grant and Johnson stated that this result demands further confirmation (*73*). It is probable that the acidic conditions used caused a rearrangement (see Sect. 6.3.5) rather than reduction.

Reductive acetylation of pristimerin (**11**) with zinc-dust and Ac$_2$O either in the presence of triethylamine (*118*) or NaOAc and glacial HOAc (*73*) afforded the same product, diacetyldihydropristimerin. Although this product was thought to be identical with diacetylpristimerol (**85**) [obtained by acetylation (Ac$_2$O/pyridine) of pristimerol (**84**)], Nakanishi *et al.* provided spectroscopic evidence to prove that these two products differed (*119*) and commented that "the structures of reductive acetates of pristimerin still remain obscure". However, the regeneration of pristimerin on deacetylation (1% NaOH in MeOH, reflux) of its reductive acetate obtained using zinc-dust, Ac$_2$O, NaOAc and glacial HOAc (*73*) requires clarification.

Reductive acetylation of tingenone (**12**) with Zn dust and NaOAc in Ac$_2$O under reflux afforded a product which was identified as the dimeric acetate (**86**) (*9*). The ^{1}H-NMR spectrum of **86** in the low field region

84 R = H
85 R = Ac

86 R$_1$ = H ; R$_2$ = O
87 R$_1$ = CO$_2$Me ; R$_2$ = H$_2$
88 R$_1$ = CO$_2$H ; R$_2$ = H$_2$

compared well with the spectra of the reductive acetates obtained from pristimerin and celastrol. Thus, THOMSON and coworkers suggested that the structures of pristimerin and celastrol reductive acetates should be revised to **87** and **88**, respectively (9). The physical and spectral data of the reductive acetates prepared from tingenone and maitenin showed that they were identical (20). Reductive acetylation of pristimerin leucotriacetate (**89**) (see Sect. 6.3.4) and under the conditions used for pristimerin (see above) resulted in diacetylisopristimerin-III (**90**) (major) and diacetylpristimerol (**85**) (minor) (Scheme 9) (82). The observed product distribution suggests that the quinonemethide system of pristimerin is more prone to reduction than acetylation under these reaction conditions.

LiAlH$_4$ reduction of pristimerin (**11**) has been reported by two groups. SESHADRI's group obtained two isomeric products which were not characterized (134), whereas JOHNSON's group obtained only one product which was characterized as the corresponding hydroxycatechol (**92**; Scheme 10) (95). Reduction with LiAlH$_4$ has proved useful in establishing the relationship between dispermoquinone (**36**) and pristimerin (**11**) (112). Thus, the tetrol (**91**) produced by LiAlH$_4$ reduction of dispermoquinone underwent dehydration in situ affording a product which was identical with the hydroxycatechol (**92**) obtained by the LiAlH$_4$ reduction of pristimerin (Scheme 10). Aerial oxidation of **92** gave 21-deoxyexcelsine (**50**) whose physical and spectral data (Table 11) were identical with those obtained by JOHNSON and his coworkers (95).

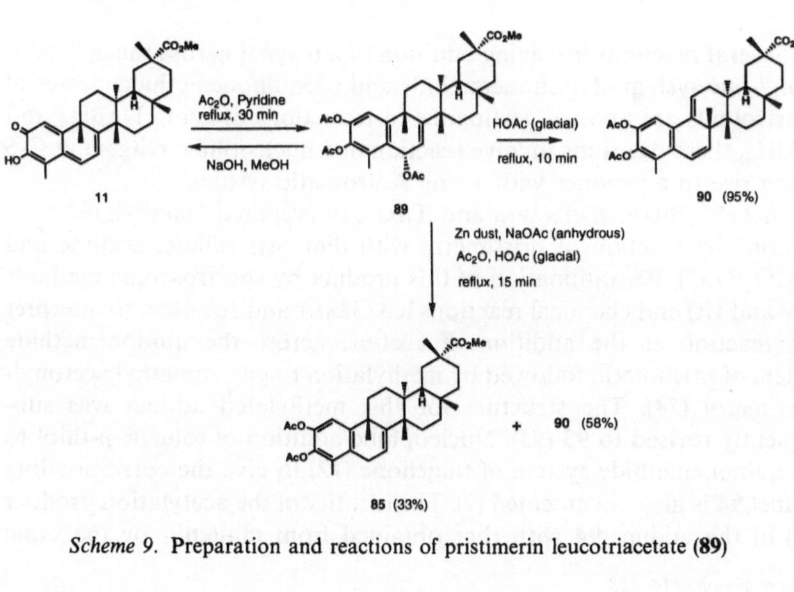

Scheme 9. Preparation and reactions of pristimerin leucotriacetate (**89**)

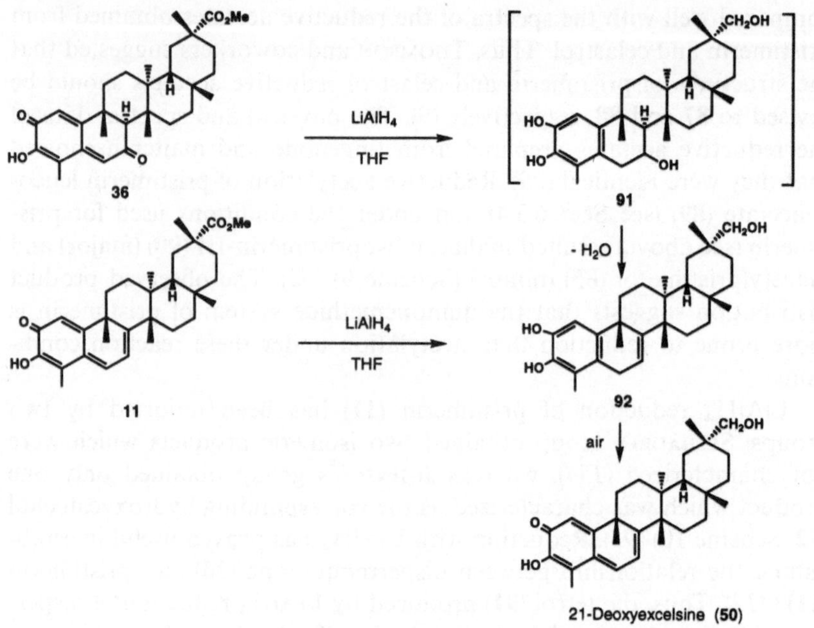

Scheme 10. Structural relationship between dispermoquinone (36) and pristimerin (11)

6.3.4. Addition Reactions

Several reactions involving addition of a reagent across the 3(4), 5(6)-dien-2-one system of quinonemethide and enequinonemethide classes of celastroloids are known. Analogous to reductions utilizing $NaBH_4$ and $LiAlH_4$, these reactions involve reaction of a nucleophilic reagent at C-6 giving rise to a product with a ring A aromatic system.

In 1955, SHAH, KULKARNI and THAKORE prepared "methylated pristimerin" by reaction of pristimerin with dimethyl sulfate, acetone and K_2CO_3 (135). Reexamination of this product by spectroscopic methods (UV and IR) and chemical reactions led GRANT and JOHNSON to interpret this reaction as the addition of acetone across the quinonemethide system of pristimerin followed by methylation to give dimethyl acetonyl-pristimerol (74). The structure for this methylated adduct was subsequently revised to 93 (95). Nucleophilic addition of toluene-p-thiol to the quinonemethide system of tingenone (12) to give the corresponding adduct 94 is also documented (9). The identity of the acetylation product (95) of the adduct 94 with that obtained from maitenin by the same

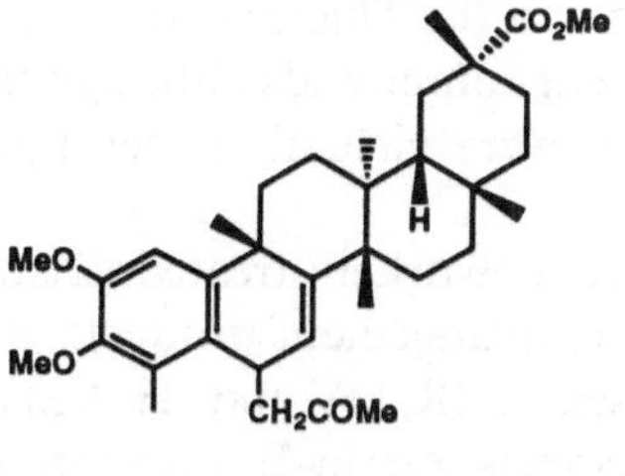

93

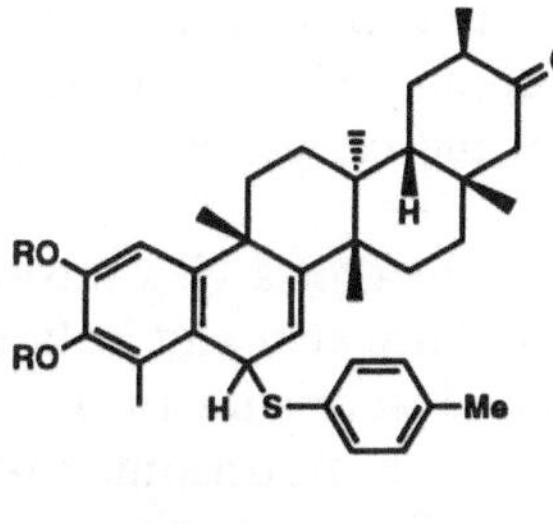

94 R = H
95 R = Ac

sequence of reactions has been used to confirm that tingenone and maitenin are identical (20).

Addition of a molecule of Ac_2O across the quinonemethide system of pristimerin occurred when pristimerin was heated with an excess of Ac_2O in the presence of pyridine (73, 74, 82). The product, pristimerin leucotriacetate, was characterized as **89** (see Scheme 9) (82). Treatment of **89** with methanolic NaOH caused regeneration of pristimerin (**11**). Although they were not characterized, pristimerin is also known to give addition products with bromine (98), sodium bisulfite (98), desoxycholic acid (98) and hydrogen chloride (74).

MARINI-BETTOLO and coworkers recently found that sodium deoxycholate formed aggregates with tingenone (**13**) which aid in solubilizing the latter in water (13). The optical behavior of tingenone was found to depend on the amount of bile acid present and the results of this optical study suggested a non-equivalent binding of tingenone to sodium deoxycholate aggregates.

6.3.5. Rearrangements

Although the quinonemethide system is quite stable under neutral conditions, it has a propensity to undergo various rearrangements under acidic conditions. In many instances rearrangement reactions have been purposely attempted during the structure elucidation of celastroloids. The rearrangement of pristimerin in hot dilute sulfuric acid was first observed by GRANT and JOHNSON in 1957 (73). The two products formed were separated by counter current distribution. An acid catalyzed rearrangement (followed by acetylation) was also observed when pristimerin was treated with sulfuric acid and acetic anhydride under Thiele conditions (73). However, a direct comparison of the major dihydroxynaphthalene produced by acid rearrangement with the dihydroxynaphthalene obtained by hydrolysis of the Thiele acetylation product later confirmed that they were different compounds although the UV spectra of the corresponding diacetates were virtually superimposable (*75*).

In the late 1960's NAKANISHI and his coworkers treated pristimerin with 2N sulfuric acid and isolated two rearrangement products, isopristimerin-I and isopristimerin-II (*88*). Based on IR, UV and ^{1}H-NMR data the structures of isopristimerin-I and isopristimerin-II were formulated as (**96**) and (**97**), respectively (see Scheme 11). A third acid catalyzed rearrangement product was also obtained when pristimerin was treated with methanolic sulfuric acid (*95*). This product, which was named isopristimerin-III, was shown to be a substituted styrene (**27**) rather than a naphthalene derivative like isopristimerin-I (**96**) and isopristimerin-II (**97**). The structures proposed for these three acid-catalyzed rearrangement products were later confirmed by MS analysis (*119*).

During the early structural investigations of pristimerin, several authors were misled by the drastic change in the IR absorption frequency of the ester carbonyl group of isopristimerin-I (1694 cm^{-1}) compared with that of pristimerin (1730 cm^{-1}) and its derivatives. This exceptional behavior of isopristimerin-I (**96**) was subsequently shown to be caused by a unique intramolecular hydrogen bonding between the methoxycarbonyl group and one of the hydroxyl groups of the ring A naphthalene diol moiety (Fig. 24) (*119*).

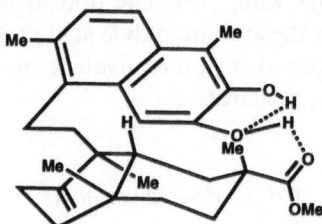

Fig. 24. Structure of isopristimerin-I (**96**) showing intramolecular H-bonding between CO_2Me and phenolic OH

Of the three isomers resulting from the acid-catalyzed rearrangement of pristimerin, isopristimerin-II may be regarded as the most stable product. It has been stated that the other two isomers, isopristimerin-I and -III, are not intermediates in the formation of isopristimerin-II (*95*). However, JOHNSON and coworkers have shown that by using perchloric acid, isopristimerin-III (**27**) can be converted into isopristimerin-II (**97**). A reasonable mechanism for the formation of isopristimerin-I, -II and -III by acid catalyzed rearrangement of pristimerin (**11**) is depicted in Scheme 11.

References, pp. 114–123

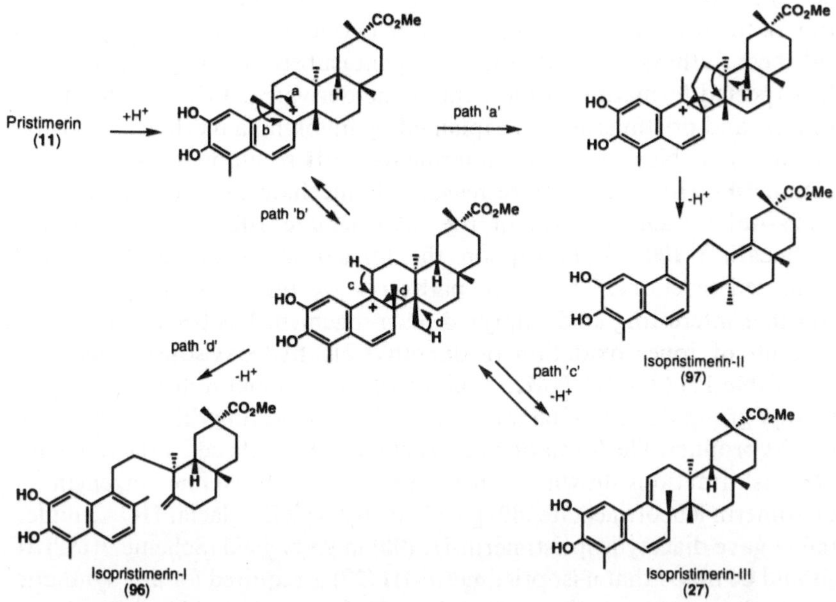

Scheme 11. Mechanism of acid catalyzed rearrangement of pristimerin (**11**)

Product analysis of the acid catalyzed rearrangement of tingenone (**12**) and 21β-hydroxypristimerin (**4**) has also been used for structure elucidation of these quinonemethide triterpenoids (*24*). When tingenone (**12**) was subjected to acid treatment under the conditions employed for pristimerin [refluxing 2N H_2SO_4 or boiling methanolic H_2SO_4] surprisingly, only one isomer, namely isotingenone-III (**28**) was produced (*126*). Unlike pristimerin, tingenone has afforded three products under Thiele conditions (Ac_2O/H_2SO_4), two of which (**98** and **99**) were derived from isotingenone-III (**28**) and the third product was identified as the diacetyl

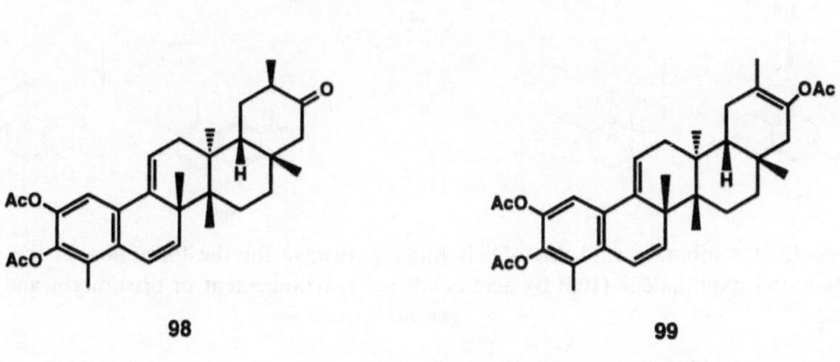

98

99

derivative of 1,2,5-trimethyl-6,7-dihydroxynaphthalene (**103**). Reinvestigation of the rearrangement of pristimerin (**11**) under Thiele conditions has led these authors to isolate the same pentasubstituted napthalene (*126*). Unexpected formation of the napthalene derivative **103**, both from tingenone and pristimerin, was explained by invoking a mechanism involving an untenable cyclobutane intermediate **101** resulting from the carbocation **100**. However, a more reasonable mechanism involving a *retro*-Diels-Alder fragmentation of the intermediate **102** formed from the carbocation **100** would explain the formation of **103** in these acid catalyzed reactions. These two pathways are presented in Scheme 12. Another interesting acid catalyzed rearrangement has been observed as a result of Jones oxidation of dimethyl-21β-hydroxyisopristimerin-III (**66**; Table 11) (*24*). ^{1}H-NMR analysis of the product indicated that the methyl group on C-8 of **66** migrated to C-7 suggesting the structure **104** for the product. The formation of **104** can be rationalized by the sequence of transformations shown in Scheme 13. Acid catalyzed rearrangement of pristimerin leucotriacetate (**89**) [see Sect. 6.3.4] with glacial HOAc under reflux gave diacetylisopristimerin-III (**90**) in 95% yield (Scheme 9) (*82*). It should be noted that if isopristimerin-III (**27**) is required for any synthetic manipulations, this is the preferred method since the acid catalyzed and

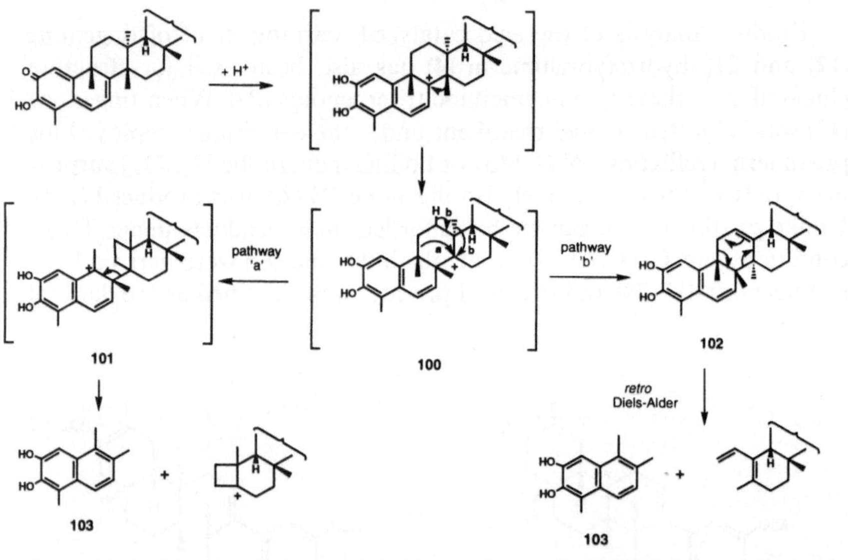

Scheme 12. Cyclobutane and *retro*-Diels-Alder pathways for the formation of the pentasubstituted naphthalene (**103**) by acid-catalyzed rearrangement of pristimerin and tingenone

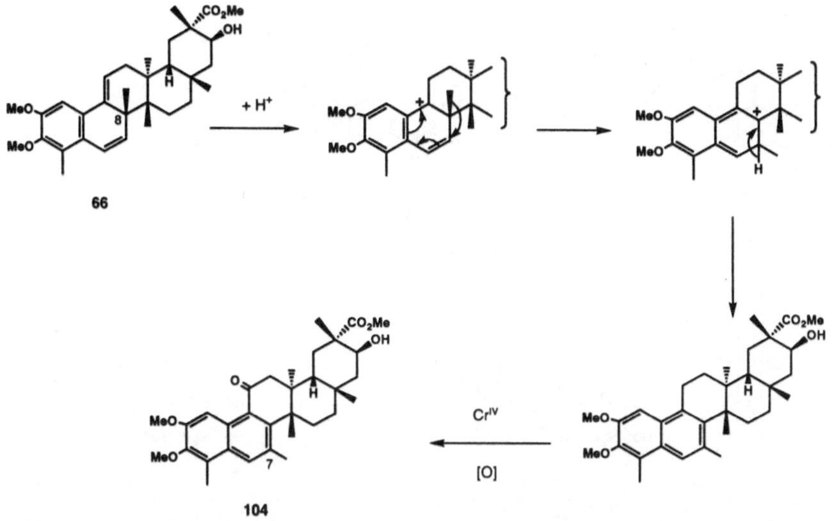

Scheme 13. Rearrangement of dimethyl-21 β-hydroxyisopristimerin-III (**66**) during Jones oxidation

photochemical rearrangement of pristimerin provides low yields of this product.

The 9(11)-enequinonemethide triterpenoid, pristimerinene (**25**), when heated in methanol containing a trace of 2N H_2SO_4, gave rise to two isomeric products which were isolated as their methyl ethers (*24*). ^{1}H-NMR data suggested that the major product was not the expected vinylnaphthalene **106**, but the divinylbenzene (**105**) related to isopristimerin-III (**27**). The mechanism involved in the formation of these two products is shown in Scheme 14.

We (*37*) and others (*65, 104*) have questioned the natural occurrence of 9(11)-enequinonemethides and suggested that their structures should be revised to the corresponding 14(15)-enequinonemethides on the basis of the stability of the isomeric aromatic (vinyl naphthalene) structure compared with the 9(11)-enequinonemethide structure (see Sect. 10). Formation of a divinylbenzene and not a vinylnaphthalene as the major product in the above acid catalyzed rearrangement further supports our view. The observed formation of a divinylbenzene (probably having structure **107** isomeric with **105**) as the major product in the acid catalyzed rearrangement can be rationalized by starting from a 14(15)-enequinonemethide [netzahualcoyene (**19**)] rather than from a 9(11)-enequinonemethide [pristimerinene (**25**)] (see Scheme 15). The reported spectral data for the divinylbenzene are not sufficient enough to differen-

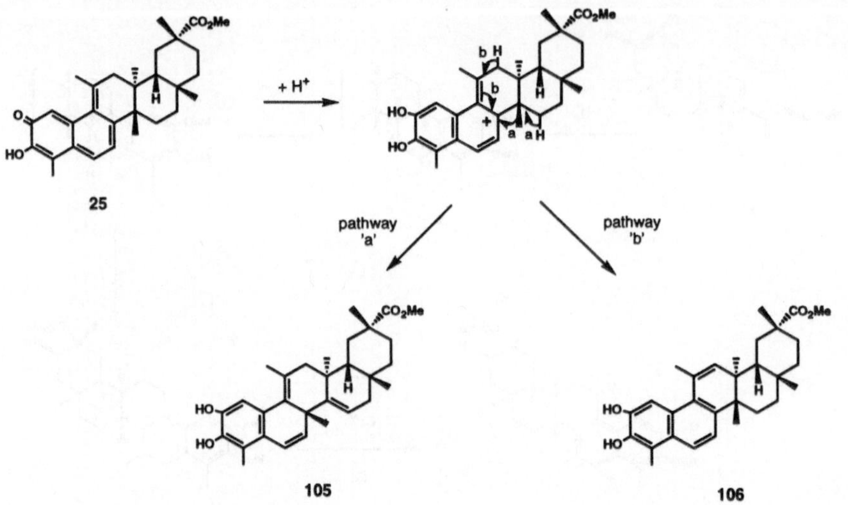

Scheme 14. Acid catalyzed rearrangement of pristimerinene (25)

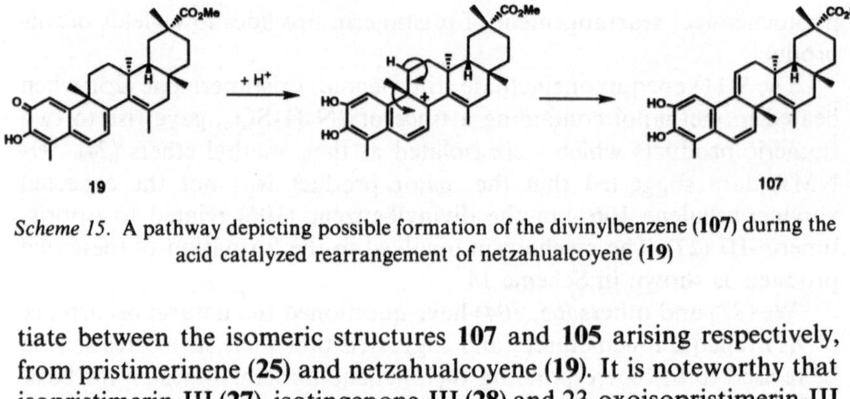

Scheme 15. A pathway depicting possible formation of the divinylbenzene (107) during the acid catalyzed rearrangement of netzahualcoyene (19)

tiate between the isomeric structures 107 and 105 arising respectively, from pristimerinene (25) and netzahualcoyene (19). It is noteworthy that isopristimerin-III (27), isotingenone-III (28) and 23-oxoisopristimerin-III (29) have been encountered as natural products suggesting the probable existence of biogenetic pathways analogous to above acid catalyzed rearrangements.

6.3.6. Photochemistry

Upon heating under reflux and irradiation with UV light for 7 days, a solution of pristimerin (11) in 95% ethanol in a quartz vessel has

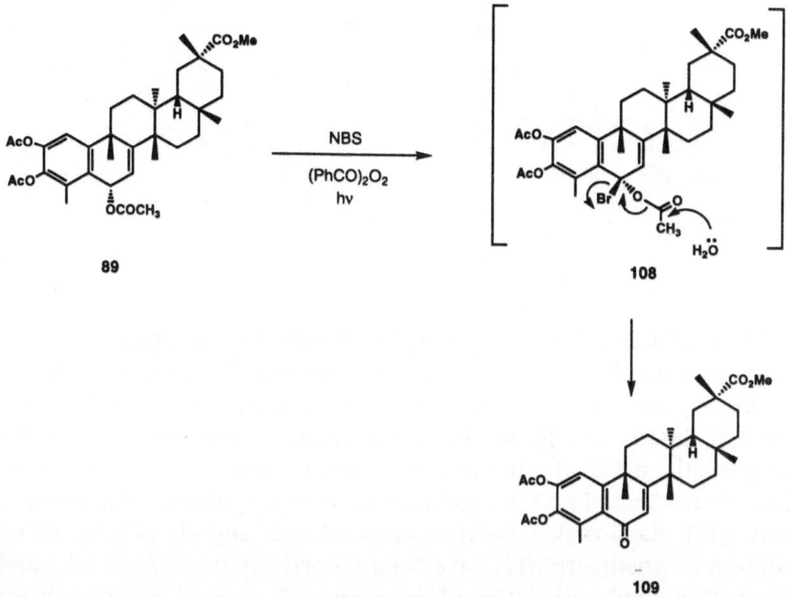

Scheme 16. Oxidation of pristimerin leucotriacetate (**89**) with NBS under photochemical irradiation

yielded isopristimerin-III (**27**) as the sole product (*95*). In a control experiment, no product formation was observed in the absence of UV irradiation.

Photochemical irradiation of a mixture of pristimerin leucotriacetate (**89**) and *N*-bromosuccinimide (NBS) in the presence of dibenzoyl peroxide as a radical initiator afforded diacetyl-6-oxopristimerol (**109**) (*82*). As shown in Scheme 16, this oxidative deacetylation of **89** may occur via an α-bromoacetate intermediate **108**.

7. Partial Synthesis

Although no celastroloid has so far been prepared by total synthesis, a total synthesis of friedelin (**110**), the likely biosynthetic precursor of celastroloids (Sect. 8) has been achieved (*91*) and the chemical conversion of friedelin into **111**, a more advanced biosynthetic precursor of celastroloids, has also been reported (*68*).

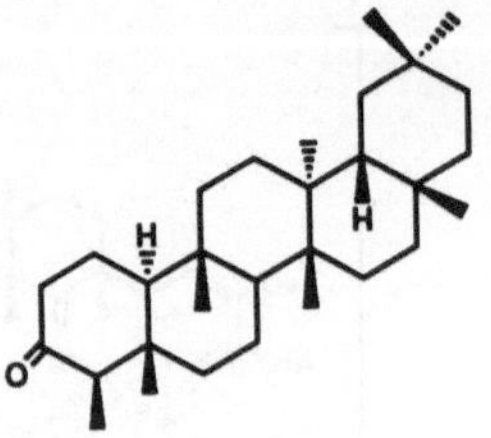

110 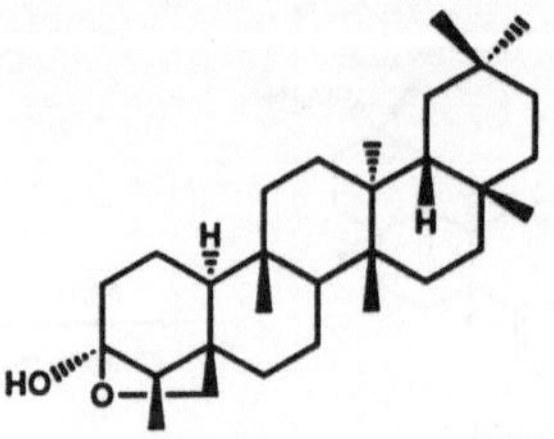   111

The partial synthesis of celastroloids has been attempted for the purpose of unambiguous structure confirmation. Thus, the structure of trimethylzeylasterone (73) derived from the natural 6-oxophenolic triterpenoid zeylasterone (35) has been confirmed by comparison with the semisynthetic material obtained from pristimerin (11) by the routes depicted in Scheme 17. One step biomimetic type syntheses of netzahualcoyene (19), Rzedowskia bistriterpenoids-I (42) and -II (43) by DDQ oxidation of pristimerin (11) have been reported (69) (see Sect. 6.3.2 and Scheme 8). A similar oxidation of tingenone (12) resulted in a convenient semisynthesis of the dimeric celastroloids, umbellatin α (44) and umbellatin β (45) (72).

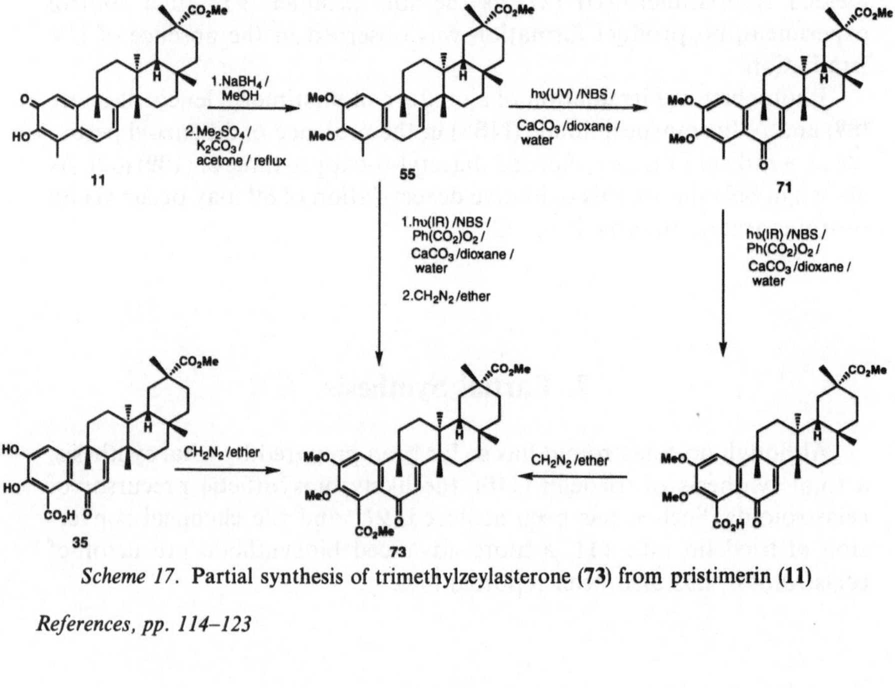

Scheme 17. Partial synthesis of trimethylzeylasterone (73) from pristimerin (11)

8. Biosynthetic Aspects

Although no experimental proof has thus far been provided for the biosynthetic origin of celastroloids, their co-occurrence with friedelanes in several plants and in cell cultures has prompted postulation of biosynthetic pathways implicating polpunonic acid (112) (104, 110), zeylanol (113) (78, 83), salaspermic acid (114) (78, 145) and orthosphenic acid (115) (59) as possible precursors. A biosynthetic pathway combining these postulates is depicted in Scheme 18. Incorporated into this pathway is cangoronine (116) (93) structurally related to salacenonal (117) recently encountered in *Salacia reticulata* var. β-diandra (141), a plant containing a variety of friedelanes and celastroloids (27).

The co-occurrence of pristimerin (11), zeylasteral (34), zeylasterone (35) and 23-oxoisopristimerin-III (29) in *Kokoona zeylanica* has led to the postulation of a biogenetic relationship between quinonemethide, phenolic and 6-oxophenolic triterpenoids (41, 78) (Scheme 19). As depicted in Scheme 20, the possible biogenetic conversion of quinonemethides, [e.g. 22β-hydroxytingenone (8)] into 14(15)-enequinonemethides, [e.g.

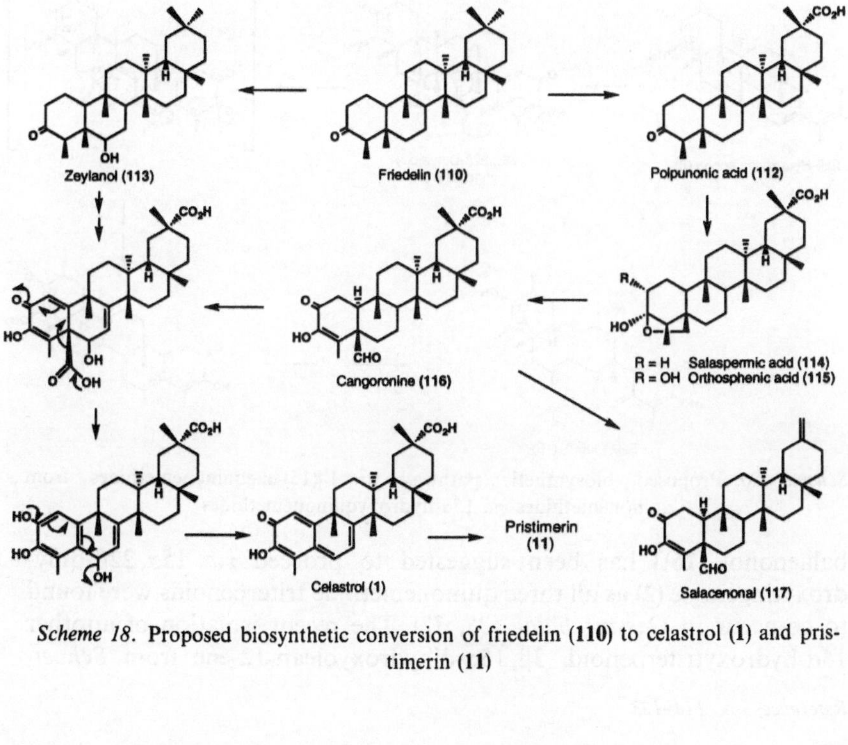

Scheme 18. Proposed biosynthetic conversion of friedelin (110) to celastrol (1) and pristimerin (11)

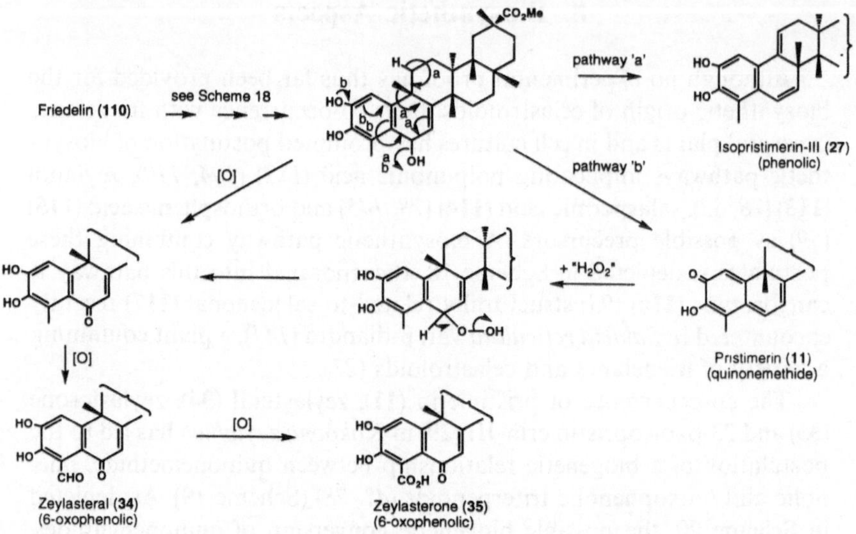

Scheme 19. Probable biogenetic relationship between quinonemethides, phenolic and 6-oxophenolic triterpenoids

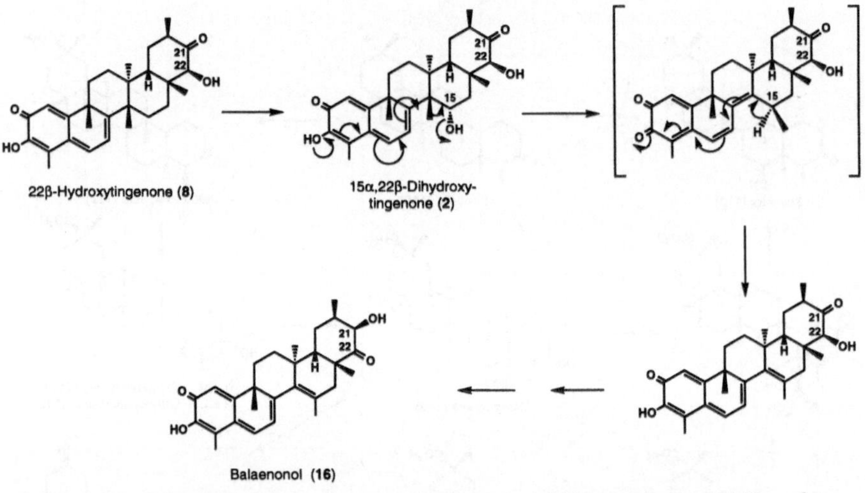

Scheme 20. Proposed biosynthetic pathway to 14(15)-enequinonemethides from quinonemethides *via* 15α-hydroxyquinonemethides

balaenonol **(16)**] has been suggested to proceed *via* 15α, 22β-dihydroxytingenone **(2)** as all three quinonemethide triterpenoids were found to co-occur in *Cassine balae* (*28, 37*). The recent isolation of another 15α-hydroxytriterpenoid, 3β, 15α-dihydroxyolean-12-ene from *Schaef-*

feria cuneifolia of the family Celastraceae is also of biogenetic significance (*71*).

Unlike their biosynthetic congeners, the celastroloids contain a variety of oxygenated substituents at C-21, C-22, C-28 and C-29. As noted previously, in all natural celastroloids C-29 is either oxidized or deleted as a result of an oxidative decarboxylation. The natural occurrence of a variety of ring E functionalized celastroloids has prompted MARINI-BETTOLO, THOMSON and coworkers to postulate probable biogenetic transformations in ring E starting from celastrol (**1**) (*24*). Probable biogenetic transformations in ring E of 14(15)-enequinonemethide triterpenoids have also been postulated to explain the biogenetic origin of this class of triterpenoids in *Cassine balae* (*37*). A combined version of these two pathways incorporating several recently encountered ring E functionalized celastroloids is given in Scheme 21. The recent isolation

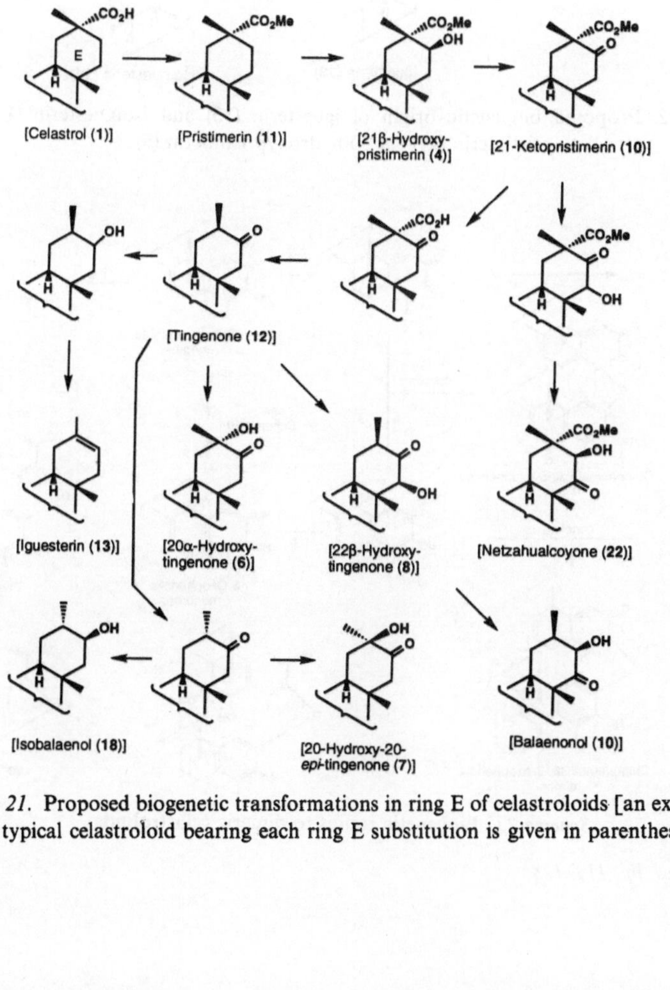

Scheme 21. Proposed biogenetic transformations in ring E of celastroloids [an example of a typical celastroloid bearing each ring E substitution is given in parenthesis]

of 30-hydroxypristimerin (5) from *Salacia reticulata* var. β-diandra (27) in which the occurrence of isoiguesterin (14) has previously been reported (142) suggests a probable biosynthetic relationship between pristimerin (11), 30-hydroxypristimerin (5), isoiguesterin (14) and iguesterin (13) as depicted in Scheme 22.

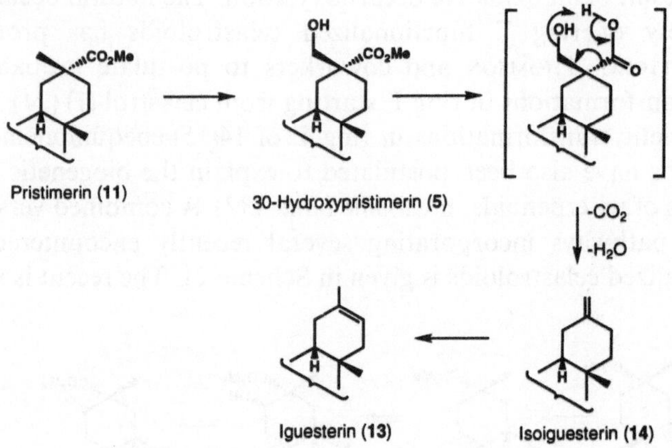

Scheme 22. Proposed biogenetic origin of iguesterin (13) and isoiguesterin (14) from pristimerin (11) *via* 30-hydroxypristimerin (5)

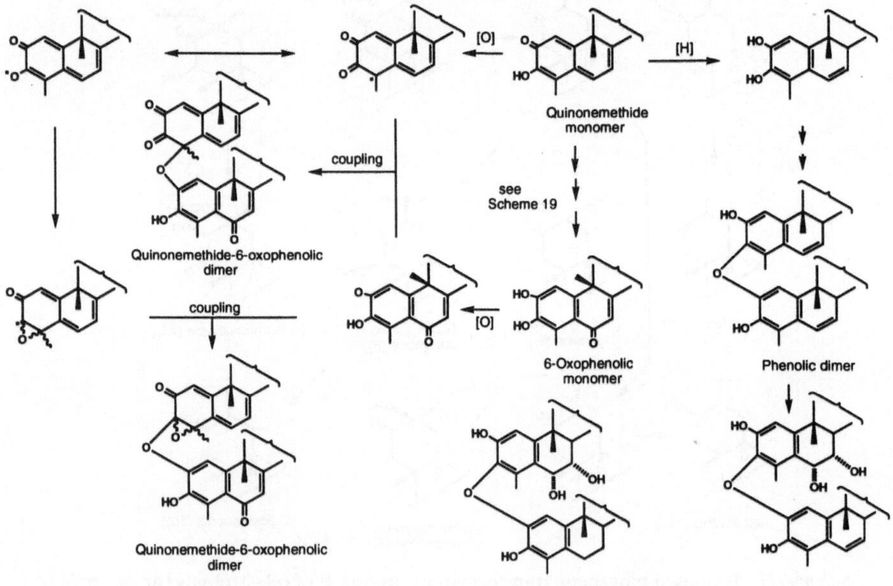

Scheme 23. Biogenetic routes to dimeric celastroids

References, pp. 114–123

The presence of at least one phenolic ring A in all naturally occurring dimeric celastroloids encountered thus far suggests that these may arise by oxidative phenolic coupling of monomeric celastroloids. Depicted in Scheme 23 are possible biogenetic routes to all known types of dimeric celastroloids starting from a monomeric quinonemethide precursor. Co-occurrence of these monomers in plants from which dimeric celastroloids have been isolated (*69, 72, 92*) coupled with the chemical evidence for formation of quinonemethide-6-oxophenolic dimers as major products during DDQ oxidation of quinonemethide triterpenoids (*69, 72*) (see Sect. 6.3.2 and Scheme 8) further supports the proposed biogenetic pathway.

9. Biological Activity

The family Celastraceae contains several species with claims that they are useful medicinally. BRÜNING and WAGNER listed some of these together with possible class(es) of compound(s) responsible for their biolog-

Plate 1. A photograph showing "Kokum soap cakes". Each soap cake is about 7 cm in diameter and 1.25 cm in thickness

ical activity (*10*). Celastroloids from *Celastrus* species have been listed as having antileukemic and antitumor properties. *Catha edulis*, the khat tree, is cultivated in the Middle East and Ethiopia for its leaves used to chew or prepare Arabian tea and Ethiopian honey wine (*4, 148*). Extracts from *Euonymus americanus*, *E. purpureus*, *Elaeodendron glaucum*, *Maytenus boaria*, *M. ilicifolia*, *M. senegalensis* and *Hippocratea capulcensis* find uses in North American and other native medicines (*148*). In Sri Lanka, the outer stem bark of *Kokoona zeylanica* is powdered and used in traditional medicine as a snuff to relieve headaches. A paste prepared by mixing powdered yellow bark of *K. zeylanica* with water is dried and formed into flat pieces to produce "kokum soap" (Plate 1) and this soap is used by some villagers of Sri Lanka in place of toilet soap. Chemical investigation of "kokum soap" has resulted in the isolation of several celastroloids, some with known antiseptic and anticancer properties (*78*).

9.1. Antimicrobial Activity

The claim that roots of *Pristimeria indica* were effective against respiratory diseases prompted BHATNAGAR and DIVEKAR to search for the active ingredient by a bioactivity-guided approach. This resulted in the isolation of pristimerin (**11**) in 1951 (*6, 7, 8*). Pristimerin exhibited considerable *in vivo* activity against a large number of gram positive cocci, particularly against *Streptococcus viridens*, the causal agent of sore throat, tonsillitis, streptococcal arthritis and against *Streptococcus faecalis*, which is known to cause urinary complications. Other gram positive organisms whose growth was inhibited by pristimerin included *Streptococcus pyrogenes*, *Staphylococcus aureus*, and *Diplococcus pneumoniae* (types I and II). Interestingly, no anti-bacterial activity was observed against any of the gram negative organisms (*Escherichia coli*, *Klebsiella pneumoniae*, *Proteus vulgaris*, *Salmonella typhii*, *S. paratyphii*, *Shigella dysenteriae* and *Vibrio cholerae*) tested. Encouraged by the preliminary *in vivo* antibiotic activity shown by pristimerin, BHATNAGAR and DIVEKAR proceeded with acute and chronic toxicity evaluation of the drug in mice followed by clinical trials on human subjects. In acute toxicity studies, toxic manifestations were observed when pristimerin was administered parenterally. Thus the chronic toxicity studies were restricted to the oral route, which was also the therapeutic route in *in vivo* experiments. Twenty mice were given 0.5 g/kg (one-fifth of the total therapeutic dose) of the drug daily for a period of 28 days and no signs of toxicity were noticed either during the period of administration of the drug or long thereafter. Quite interestingly, even before the structure was known, pristimerin (**11**) was clinically evaluated by these workers on 19 patients with low grade infection of the nasopharyngeal mucosa which did not respond to penicillin or sulpha drugs available at the time. When the drug was given orally at a dose of 20 mg twice a day for 2–3 days, recession in the size of the tonsils, amelioration of general malaise and disappearance of temperature have been observed. α-Hemolytic *Streptococci* belonging to the viridans group which were isolated from the throat swabs of the patients prior to treatment were found to be absent after the treatment with pristimerin (*6*). This pioneering work of BHATNAGAR and DIVEKAR encouraged subsequent workers to search for bioactive compounds in Celastraceae species.

In a search for antimicrobial agents in higher plants, the Brazilian group headed by GONCALVES DE LIMA compared the antibiotic activity of celastrol (**1**) with that of pristimerin (**11**) and found that pristimerin showed lower antibiotic activity than celastrol (*53*). Like pristimerin, celastrol exhibited inhibitory effects against several gram positive bacteria, especially *Bacillus subtilis, B. mycoides* (= *B. cereus mycoides*), *B. anthracis, Staphylococcus aureus* W., *Streptococcus faecalis, Streptococcus pyrogenes, Brucella abortus* and *Mycobacterium phlei*. In continuing their search for antimicrobial agents the same group of workers isolated maytenin [tingenone (**12**)] from *Maytenus ilicifolia* and found it to have *in vitro* inhibitory activity against microorganisms such as *Bacillus subtilis, Staphylococcus aureus, Sarcina lutea, Streptococcus faecalis, S. haemolyticus, Mycobacterium smegmatis,* and *Gibberella fujikuroi* (*54*).

In a structure-activity relationship (SAR) study of the antimicrobial activity of quinonemethide and 14(15)-enequinonemethide triterpenoids, GONZALEZ and his coworkers evaluated such celastroloids isolated from *Schaefferia cuneifolia* and *Maytenus horrida* as well as their derivatives (*115*). The study included the quinonemethides, celastrol (**1**), pristimerin (**11**) and tingenone (**12**), the 14(15)-enequinonemethides, netzahualcoyene (**19**), netzahualcoyondiol (**21**), netzahualcoyone (**22**), and netzahualcoyonol (**23**), the reductive triacetate (**118**) derived from **21** and two nonquinonemethide triterpenoids, polpunonic acid (**112**) and salaspermic acid (**114**). Netzahualcoyone (**22**) showed strong activity against the bacteria, *Staphylococcus aureus, S. epidermidis, S. warneri, S. saprophyticus, Bacillus subtilis, B. cereus, B. megaterium, B. pumilus, B. alvei* and *Micrococcus luteus,* but was found to be inactive against *Escherichia coli, Pseudomonas aeruginosa, Enterobacter hafniae, Salmonella* sp., *Proteus mirabilis* and *Klebsiella pneumoniae*; netzahualcoyone also showed moderate activity against the yeast, *Saccharomyces cerevisiae*. The MIC (minimum inhibitory concentration) of netzahualcoyone for *Streptococcus aureus* (1.5 to 1.6 µg/ml) can be compared favorably with that reported for other antibiotics, while its MIC for *Bacillus subtilis* was 1.25 to

118 $R_1 = R_2 = Ac$, $R_3 = \beta$-OH,α-H
119 $R_1 = Me$, $R_2 = H$, $R_3 = O$

0.625 μg/ml, and for *Escherichia coli* it was more than 200 μg/ml. Based on this SAR study the following conclusions were made: a quinonemethide triterpene skeleton [with or without 14(15)-unsaturation] and groups in ring E are structural requirements for antimicrobial activity and the most effective group in ring E is a carboxyl group at C-20.

In an attempt to understand the mode of antimicrobial action of netzahualcoyone (22), GONZALEZ and coworkers studied its activity on *Bacillus subtilis* and *Escherichia coli* (*116*). Netzahualcoyone (22) was found to inhibit the respiration of intact cells of the gram-positive bacterium, *B. subtilis*, but had no effect on the respiration of intact cells of the gram-negative bacterium, *E. coli*. Interestingly, when preparations of sonically disrupted cells were examined, inhibitory activity on both bacteria were observed. In order to understand this difference, the effect of netzahualcoyone on the respiration of *Saccharomyces cerevisiae* and bacteria was studied. The results indicated that netzahualcoyone had an inhibitory effect on the cellular respiration of the microorganisms and were in line with previous findings by GONZALEZ and coworkers (*67*) of its effect on the growth of *S. cerevisiae* and its MIC's for *B. subtilis* and *E. coli*. These results led the authors to the important conclusion that the insensitivity of the intact gram negative bacteria to netzahualcoyone (22) may be due to the existence of a permeability barrier, probably the outer membrane of these bacteria (*116*).

In a recent study GONZALEZ *et al.* evaluated the antimicrobial activity of four novel dimeric quinonemethide-6-oxophenolic triterpenoids, Rzedowskia bistriterpenoid-I (42), Rzedowskia bistriterpenoid-II (43), umbellatin α (44) and umbellatin β (45), on the bacteria *Staphylococcus aureus, Bacillus subtilis, B. cereus, Salmonella* sp. and *Escherichia coli* (*72*). As for other celastroloids, the gram positive bacteria were found to be more susceptible to the activity of these compounds with the exception of umbellatin β, which was devoid of activity. The most active was found to be the Rzedowskia bistriterpenoid-II (43) with an MIC of 1–2 μg/ml on *B. subtilis*. The dimers were active only on *B. subtilis* and even then were found to be less active than their quinonemethide monomers, pristimerin (11) and tingenone (12).

9.2. Antitumor Activity

In the early 1960's, prior to the availability of tumor cell lines, SCHWENK evaluated the antitumor potential of celastrol (1), pristimerin (11) and pristimerol (84), as part of a study of the tumor growth inhibiting properties of a variety of quinones in the golden hamster cheekpouch test (133). For celastrol, 22–57% tumor growth inhibition was observed whereas for pristimerin and pristimerol the inhibitions were 31–39% and 16%, respectively. In an attempt to correlate the observations made in his study with a possible mechanism of action of quinonoid substances in tumor chemotherapy, SCHWENK postulated that the reduction of quinonoids forms hydroquinones which by autooxidation give H_2O_2 and that H_2O_2 destroys tumor cells by inhibition of their glycolysis.

Pristimerin (11) was later evaluated for antitumor activity in *in vivo* studies and was found to inhibit growth of Yoshida and 180 sarcomas and Ehrlich carcinomas at a dose of 50 mg/kg/day (51). Encouraged by these results, FERREIRA DE SANTANA and coworkers subjected maytenin [tingenone (12)] to a clinical study involving 25 patients having various types of neoplasia in advanced phase and resistant to other available anticancer drugs (38). The results with maytenin indicated that it was somewhat effective against epidermoid carcinoma of the tonsil pillars, uterus epidermoid carcinoma, stomach carcinoma and lymphoepithelioma. In another clinical study 11 patients bearing basic cellular carcinomas were treated with maytenin [tingenone (12)], the latter was less active than primin and plumbagin, two other natural quinones tested (114). However, maytenin had comparatively low irritant action and late antineoplastic properties. In order to understand the mechanism of antitumor action of tingenone (12), ANGELETTI and MARINI-BETTOLO studied the effect of this quinonemethide on RNA and protein synthesis in cell cultures (2). Their results indicated that even at low concentrations tingenone exerted a marked inhibitory effect on *in vitro* protein and RNA synthesis.

With the availability of more celastroloids and their derivatives, GONZALEZ and his coworkers (58) extended their studies to pristimerin (11), tingenone (12), iguesterin (13) and their diacetyldihydro derivatives 85, 52 and 58 (Table 11). HeLa 229 and Ehrlich ascitic (mice) cell lines were used to evaluate their cytostatic action and effects on DNA, RNA and protein synthesis, respectively. All six compounds showed powerful *in vitro* cytostatic activity in median inhibitory doses (ID_{50}) of between

0.3 and 0.6 µg/ml, but the most active were tingenone (12) and diacetyl-pristimerol (85) with activities slightly inferior to the positive control used, 6-mercaptopurine (ID$_{50}$ = 0.1 µg/ml). Although the results did not provide any information on structure-activity relationships, it was found that all six compounds stopped the formation of DNA, RNA and protein synthesis within a few minutes. In a more detailed study ANGELETTI and MARINI-BETTOLO tested maitenin [tingenone (12)] for its *in vitro* effect on the rate of incorporation of the base uridine into RNA and of amino acids in proteins (2). Their results confirmed the previous findings that tingenone exerted a marked inhibitory effect on RNA and protein synthesis. GONZALEZ *et al.* have recently evaluated the 14(15)-enequinonemethides, netzahualcoyene (19), netzahualcoyondiol (21), and netzahualcoyone (22), and one of their derivatives, dimethyldihydronet-zahualcoyone (119) (see p. 98) for cytostatic activity in HeLa cells. Their results are depicted in Table 27 along with results of some other cytotoxicity studies on a number of celastroloids. Two findings worth noting are that netzahualcoyone (22) was as active as the anticancer drug 6-mercaptopurine and that the chemical reduction of the quinonemeth-ide system caused 700 fold reduction in the cytostatic activity compared with the parent 14(15)-enequinonemethides (66).

The Chicago group recently evaluated the *in vitro* cytotoxic activities of celastrol (1), 20-hydroxy-20-*epi*-tingenone (7), 22β-hydroxytingenone (8) and tingenone (12) on a battery of human cancer cell lines (3, 122). Their results, summarized in Table 28, suggested that although these celastroloids exhibited strong *in vitro* cytotoxic activity, none of them showed any selective activity. In an attempt to evaluate the anticancer potential of isoiguesterin (14), SNEDEN subjected this quinonemethide triterpenoid to both *in vitro* and *in vivo* cytotoxicity assays and toxicity testing (136). When tested for *in vivo* activity against the P388 lym-phocytic leukemia in mice at two dose levels (T/C 147 at 1.00 mg/kg, 125 at 0.75 mg/kg) and *in vitro* against the KB cell culture (ED$_{50}$ 0.24 mg/ml) it exhibited encouraging activity. However, isoiguesterin was demon-strated to be toxic above 1.00–2.00 mg/kg, thus overriding its therapeutic effect.

It is noteworthy that pristimerin (11) is listed as a toxic compound in Sax's Dangerous Properties of Industrial Materials (106). Pristimerin is reported to be poisonous by intraperitoneal (i.p.), subcutaneous and oral routes with LD$_{50}$'s of 200, 400 and 8000 mg/kg, respectively. In a study of pharmacodynamic and histopathological effects of maytenin [tingenone (12)] in mice, FERREIRA DE SANTANA's group found its LD$_{50}$ to be 19.39 mg/kg (i.p. route) and had no effects on erythrocytes and leucocytes in rats at a dose of 1 or 4 mg/kg/day (i.p. route) for 30 days (39).

Table 27. *Cytotoxicity of Some Celastroloids and Their Derivatives*

Compound	Cell line			
	KB	P-388	HeLa	Reference(s)
Celastrol (1)[a]	1.2[b]	3.5	–	(122)
20-Hydroxy-20-*epi*-tingenone (7)[a]	1.7[b]	2.0	–	(122)
22β-Hydroxytingenone (8)[a]	2.5[b]		–	(104)
	0.62[b]	0.073		(3)
Pristimerin (11)	0.23[c]	0.052[c]	0.6[d]	(58, 93)
Tingenone (12)[a]	0.27		0.3[d]	(58, 104)
	0.50[b]	0.1		(122)
Iguesterin (13)	–	–	0.6[d]	(58)
Isoiguesterin (14)	0.24[b]	–	–	(136)
Netzahualcoyene (19)	–	–	> 1[d]	(66)
Netzahualcoyondiol (21)	–	–	1[d]	(66)
Netzahualcoyone (22)	–	–	0.1[d]	(66)
Netzahualcoyonol (23)	–	–	> 1[d]	(66)
Isopristimerin-III (27)	1.7[c]	2.0[c]	–	(93)
Isotingenone-III (28)	1.1[c]	0.18[c]	–	(93)
Cangorosin B (41)		0.0037[e]	–	(92)
Diacetylpristimerol (85)	–	–	0.3[d]	(58)
Diacetyldihydrotingenone (52)	–	–	0.4[d]	(58)
Diacetyldihydroiguesterin (59)	–	–	0.4[d]	(58)
Dimethyldihydronetzahualcoyone (119)	–	–	70[d]	(66)

[a] For cytotoxicity data in other cell lines, see Table 28.
[b] ED_{50} (µg/ml).
[c] IC_{50} (µg/ml).
[d] ID_{50} (µg/ml).
[e] IC_{50} (mol/l).

In a study of the possible interaction of tingenone (12) with DNA, D'ALAGNI *et al.* found spectroscopic (UV and CD) evidence to support such an interaction, but they were unable to gain any evidence on the relationship between the binding mode and the base composition of DNA (*12*). They concluded that size and shape of the tingenone molecule are favorable for its inclusion in the narrow groove of DNA and that hydrogen bonds can be formed between the hydroxy group of tingenone and the phosphate groups of DNA. They further commented that "since tingenone is a bulky molecule, an intercalation mechanism seems to be unlikely, unless only the nearly planar A and B rings are engaged". Tingenone (12) also showed activity (LC_{50} 298 ppm) in the brine shrimp lethality test, a preliminary bioassay used to detect antitumor agents (*11*). Interestingly, celastrol (1), excelsine (3) and pristimerin (11) were found to

Table 28. *Cytotoxicity of Some Celastroloids in Human Cancer Cell Lines*

| Human Cancer Cell | | ED_{50} (µg/ml) | | | |
Type	Line	Celastrol (1)	20-Hydroxy-20-*epi*-tingenone (7)	22β-Hydroxy-tingenone (8)	Tingenone (12)
Breast	BC-1	0.3	0.9	0.75	0.3
Breast [Hormone dependent]	ZR-75-1	0.4	1.0	–	0.6
Colon	Col-2	2.3	5.8	0.61	0.9
Epidermoid carcinoma	A-431	0.6	1.1	–	0.7
Oral epidermoid	KB	1.2	1.7	0.62	0.5
Oral epidermoid [Multidrug-resistant]	KB-VI(+ VLB)	1.7	3.3	–	0.4
	KB-VI(– VLB)	2.4	3.0	–	0.4
Fibrosarcoma	HT-1080	0.5	2.1	0.44	0.7
Glioblastoma	U 373	0.4	0.8	–	0.2
Lung	LU-1	1.3	2.2	0.84	0.6
Melonoma	MEL-2	0.6	1.5	0.39	0.2
Prostate	LNCaP	0.3	1.0	–	0.2

be inactive even at a concentration of 1000 ppm. Evaluated in a bioassay involving the germination of corn and lentil seeds, pristimerin (11) and tingenone (12) were found to exhibit strong inhibitory activity (30).

9.3. Other Biological Activities

Besides antimicrobial and antitumor activities, celastroloids have also shown antimalarial (schizontocidal), and spermicidal activities. Claims of antimalarial activity for *Celastrus paniculatus* prompted PAVANAND and coworkers to initiate a study of this species and in 1989 they reported a bioactivity-guided isolation of pristimerin (11), the constituent responsible for its antimalarial activity (125). However, when tested *in vitro* against various multidrug resistant isolates of the malarial parasite, *Plasmodium falciparum*, pristimerin was found to be less active than the conventional antimalarial drugs tested.

It has been reported recently that some quinonemethide and 14(15)-enequinonemethide triterpenoids cause immobilization of human spermatozoa thus revealing their potential as male contraceptive agents (127). Of the 11 celastroloids tested the activity of tingenone (12), 20α-hydroxytingenone (6) [more correctly 20-hydroxy-20-*epi*-tingenone (7) (107)], 22β-hydroxytingenone (8), isoiguesterol [30-hydroxy-20(30)-dihydroisoiguesterin (9)] and isobalaendiol (17) was found to depend on the nature and location of the functional groups in ring E. The phenolic triterpene, zeylasterone (35) had no antifertility activity in the female rat when tested using the World Health Organization MB 30 protocol (85). Pristimerin (11) has shown weak antihemorrhagic activity at a dose of 25 mg/kg whereas celastrol (1) was found to be inactive even at a dose of 100 mg/kg (49). Maytenin [tingenone (12)] was reported to have no effect on the isolated guinea pig ileum or *in situ* toad heart at a dose of 2 mg/ml (39). It also had no significant effect on blood pressure and respiration in anesthetized rats. At a dose of 10 mg/kg (i.p. route) maytenin showed no antiinflammatory effect in rats. However, irritant effects at the site of skin injections were found within 15 days in rats given 10 or 12 mg/kg/day for 5 days (39).

10. Conclusions

Celastroloids, although comparatively small in number, constitute a group of triterpenoids with unique structural features and biological activities. Their restricted occurrence in the plant family Celastraceae (including Hippocrateaceae) is of chemotaxonomic significance. Some celastroloids exhibit antitumor and antibiotic properties. Thus their dual bioactivity parallels those of some microbial metabolites.

It has been suggested recently that the structure of 20α-hydroxytingenone (**6**) should be revised to 20-hydroxy-20-*epi*-tingenone (**7**) (*107*). Structures of celastroloids, except those of the 9(11)-enequinonemethide class, have been rigorously established by the application of modern NMR and other spectroscopic techniques. Structure elucidation of the three 9(11)-enequinonemethides thus far known has relied only on UV, ^{1}H-NMR data and analysis of acid catalyzed rearrangement products, some of which are ambiguous. In 1979, MARINI-BETTOLO, THOMSON and coworkers stated (*24*) that the available evidence for the structure proposed by REDDY *et al.* (*129, 130*) for the first 9(11)-enequinonemethide to be isolated, namely salacia quinonemethide (**26**), is limited suggesting that it may be identical with pristimerinene (**25**) encountered by them. KUTNEY and his group while agreeing with MARINI-BETTOLO, THOMSON *et al.* pointed out that structure **26** proposed for salacia quinonemethide

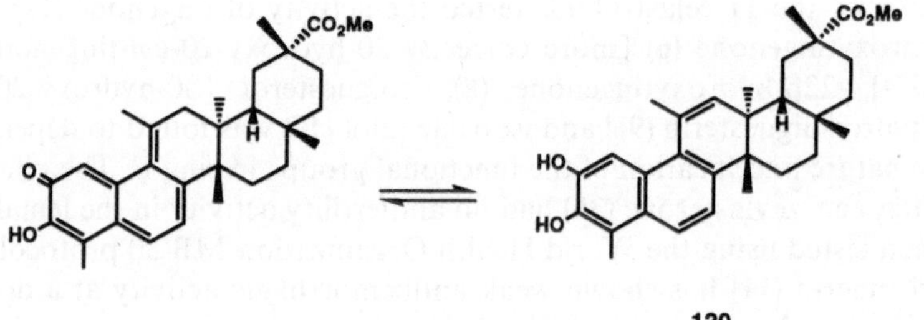

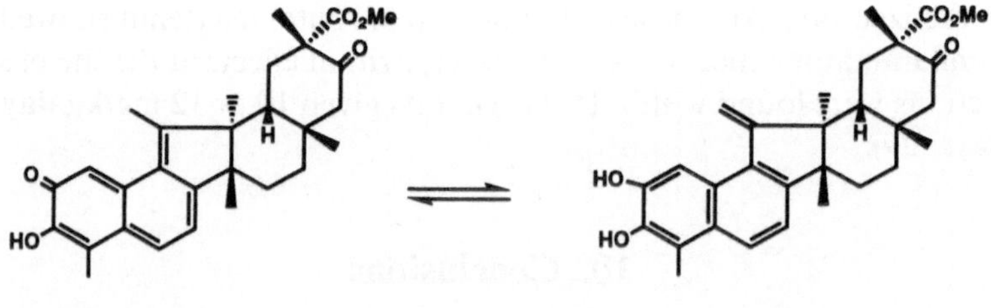

Fig. 25. Tautomeric aromatic structures for the 9(11)-enequinonemethides, pristimerinene (**25**) and salacia quinonemethide (**26**)

is unlikely as it would exist in the more stable aromatic (vinylnaphthalene) form represented by structure **121** (*104*).

The 9(11)-enequinonemethide structure proposed for balaenonol was revised to 14(15)-enequinonemethide (**16**) based on stability arguments and the results of nOe studies (*36, 37*). The possible identity of the 9(11)-enequinonemethide, pristimerinene (**25**) with the 14(15)-enequinonemethide, netzahualcoyene (**19**) has also been suggested (*65*). The formation of the divinylbenzene (**105**) as the major product during the acid catalyzed rearrangement of pristimerinene (**25**) (Schemes 14 and 15) further supports this contention (see Sect. 6.3.5). MM-2 calculations revealed that the aromatic (vinyl naphthalene) forms of 9(11)-enequinonemethides (*e.g.* **120**) are ca. 40 kcal mole^{-1} more stable than the corresponding enequinonemethide (*e.g.* **25**) form. The accumulated evidence suggests that salacia quinonemethide, pristimerinene and netzahualcoyene are one and the same compound with the 14(15)-enequinonemethide structure **19**.

Many reaction products obtained from celastrol (**1**) and pristimerin (**11**) in the course of early structural studies have yet to be fully characterized (see Sect. 6.2.1). A recent study proved to be somewhat rewarding (*82*). The accumulated data on bioactivity and clinical effectiveness warrant further biological evaluation of celastroloids and their derivatives.

Addendum

After completion of this review five additional reports on triterpenoid quinonemethides have appeared in the literature: one dealing with a unique class of quinonemethides bearing a shionane (D:A-*friedo*-18,19-secolupane) triterpenoid skeleton and the remaining four on quinonemethide, aromatic and dimeric classes of celastroloids. The occurrence of triterpenoid quinonemethides outside the family Celastraceae (and Hippocrateaceae) has been reported for the first time (*S1*). Listed in Table 29 are the sources and some physical data of the recently encountered new celastroloids.

Maytenus canariensis has afforded a new triterpenoid quinonemethide, 16β-hydroxyiguesterin (**120**), which was isolated by a combination of Sephadex LH-20 and silica gel chromatography (*S3*). The structural relationship between **120** and iguesterin (**13**) was apparent from their IR, UV, and ^{1}H-NMR data. The hydroxyl group outside the quinonemethide system was located at C-16 with the help of HMBC spectra and its stereochemical disposition was determined as β on the basis of molecular

Table 29. Recent Reports of New Celastroloids

Trivial names	Source (part)[a]	Molecular formula (MW)	$(\alpha)_D$	mp (°C)	References
16β-Hydroxyiguesterin (120)	Maytenus canariensis (rb)	$C_{28}H_{36}O_3$ (420.27)	$-100°$ (CHCl$_3$)	amorphous	(S3)
6-Oxotingenol (121)	Maytenus ilicifolia (sb)	$C_{28}H_{36}O_4$ (436.26)	$-151.8°$ (pyridine)	> 300	(S5)
6-Oxopristimerol (122)	Maytenus chuchuhuasca (sb)	$C_{30}H_{40}O_5$ (480.29)	$-80.4°$ (pyridine)	173–178	(S5)
3-Methyl-6-oxotingenol (123)	M. chuchuhuasca (sb)	$C_{29}H_{38}O_4$ (450.28)	$-93.4°$ (CHCl$_3$)	273–276	(S5)
3-Methyl-22β,23-dihydroxy-6-oxotingenol (124)	M. chuchuhuasca (sb)	$C_{29}H_{38}O_6$ (482.27)	$-60.9°$ (CHCl$_3$)	240–245	(S5)
Xuxuarine Aα (125)	M. chuchuhuasca (bk)	$C_{56}H_{70}O_7$ (854.51)	$+645.2°$ (CHCl$_3$)	amorphous	(S6)
Xuxuarine Aβ (129)	M. chuchuhuasca (bk)	$C_{56}H_{70}O_7$ (854.51)	$-512.6°$ (CHCl$_3$)	amorphous	(S6)
Xuxuarine Bα (126)	M. chuchuhuasca (bk)	$C_{56}H_{70}O_9$ (886.50)	$+647.0°$ (CHCl$_3$)	amorphous	(S6)

Xuxuarine Bβ (130)	*M. chuchuhuasca* (bk)	$C_{56}H_{70}O_9$ (886.50)	−523.3° (CHCl₃)	amorphous	(S6)
Xuxuarine Cα (127)	*M. chuchuhuasca* (bk)	$C_{56}H_{70}O_8$ (870.51)	+654.0° (CHCl₃)	amorphous	(S6)
Xuxuarine Cβ (131)	*M. chuchuhuasca* (bk)	$C_{56}H_{70}O_8$ (870.51)	−505.4° (CHCl₃)	amorphous	(S6)
Xuxuarine Dα (128)	*M. chuchuhuasca* (bk)	$C_{56}H_{70}O_8$ (870.51)	+554.2° (CHCl₃)	amorphous	(S6)
Xuxuarine Dβ (132)	*M. chuchuhuasca* (bk)	$C_{56}H_{70}O_8$ (870.51)	−517.0° (CHCl₃)	amorphous	(S6)
7,8'-Dihydroxuxuarine Aβ (133)	*M. chuchuhuasca* (bk)	$C_{56}H_{72}O_7$ (872.52)	−562.1° (CHCl₃)	amorphous	(S6)
Magellanin (134)	*Maytenus magellanica* (rob)	$C_{60}H_{82}O_6$ (898.61)	−338.8° (CHCl₃)	amorphous	(S2)

[a] bk = bark; rb = root bark; rob = root outer bark; sb = stem bark

120

mechanics calculations and ^{1}H-NMR coupling constants of the preferred conformer.

In continuing their work on South American *Maytenus* species, ITOKAWA and coworkers have obtained aromatic and dimeric celastroloids from the stem barks of *M. ilicifolia* and *M. chuchuhuasca*. *M. ilicifolia* has afforded 6-oxotingenol (**121**) (*S5*) whereas *M. chuchuhuasca* has yielded 6-oxopristimerol (**122**), 3-methyl-6-oxotingenol (**123**), and 3-methyl-22β,23-dihydroxy-6-oxotingenol (**124**) (*S5*), all of which are new aromatic celastroloids. In addition, the known quinonemethides pristimerin (**11**), tingenone (**12**) and 22β-hydroxytingenone (**8**) have been isolated from *M. chuchuhuasca*. This constitutes the second report of the occurrence of **11** and **12** in this plant (see Table 10). Structures of the new celastroloids **121**–**124** have been elucidated by a combination of ^{1}H- and ^{13}C-NMR spectral methods. The methoxy group in **123** was located at C-3 with the aid of HMBC and nOe correlation spectroscopy and the

121 $R_1 = R_3 = R_5 = H$, $R_2 = Me$, $R_4 = O$

122 $R_1 = R_5 = H$, $R_2 = Me$, $R_3 = CO_2Me$, $R_4 = H_2$

123 $R_1 = R_2 = Me$, $R_3 = R_5 = H$, $R_4 = O$

124 $R_1 = Me$, $R_2 = CH_2OH$, $R_3 = H$, $R_4 = O$, $R_5 = OH$

final confirmation of its structure performed by X-ray crystallography. All celastroloids encountered when evaluated for cytotoxicity showed moderate activities against the cultured tumor cell lines KB, P-388 and L-1210 (Table 30).

Further investigation of *M. chuchuhuasca* ("Xuxua") has afforded nine novel stereoisomeric celastroloid dimers xuxuarines Aα (**125**), Aβ (**129**), Bα (**126**), Bβ (**130**), Cα (**127**), Cβ (**131**), Dα (**128**), Dβ (**132**) and 7',8'-dihydroxuxuarine Aβ (**133**) (*S6*). All xuxuarines were composed of quinonemethide and aromatic monomers, derived from tingenone (**12**) and/or 22β-hydroxytingenone (**8**) and linked together by two ether bonds formed between the two A rings. Biogenetically these dimers can arise by an *ortho*-quinone Diels-Alder type reaction as proposed by ITOKAWA *et al.* or from the previously known dimeric quinonemethide-6-oxophenolic triterpenoids umbellatin α (**44**) and umbellatin β (**45**) (Scheme 24).

The gross structures of xuxuarines have been elucidated with the aid of EI-MS fragmentation, FAB-MS, IR, UV, ^{1}H- and ^{13}C-NMR

Table 30. *Recent Results of Some Cytotoxicity Studies on Celastroloids Using Cultured Tumor Cell Lines*

Celastroloid	Cell line			
	KB	P-388	L-1210	Reference
6-Oxotingenol (**121**)	30[a]	2.6[a]	6.0[a]	(*S5*)
6-Oxopristimerol (**122**)	2.8[a]	1.5[a]	2.8[a]	(*S5*)
3-Methyl-6-oxotingenol (**123**)	11[a]	> 100[a]	> 100[a]	(*S5*)
3-Methyl-22β,23-dihydroxy-6-oxotingenol (**124**)	–	4.3[a]	–	(*S5*)
Pristimerin (**11**)	0.55[a]	0.12[a]	0.36[a]	(*S5*)
Tingenone (**12**)	0.28[a]	0.041[a]	0.14[a]	(*S5*)
22β-Hydroxytingenone (**8**)	–	0.012[a]	–	(*S5*)
Xuxuarine Aα (**125**)	> 0.12[b]	> 0.060[b]	> 0.094[b]	(*S6*)
Xuxuarine Aβ (**129**)	> 0.12[b]	> 0.12[b]	> 0.12[b]	(*S6*)
Xuxuarine Bα (**126**)	> 0.12[b]	0.019[b]	0.020[b]	(*S6*)
Xuxuarine Bβ (**130**)	> 0.12[b]	> 0.12[b]	> 0.12[b]	(*S6*)
Xuxuarine Cα (**127**)	> 0.12[b]	0.059[b]	0.092[b]	(*S6*)
Xuxuarine Cβ (**131**)	> 0.12[b]	> 0.12[b]	> 0.12[b]	(*S6*)
Xuxuarine Dα (**128**)	> 0.12[b]	0.036[b]	0.044[b]	(*S6*)
Xuxuarine Dβ (**132**)	> 0.12[b]	> 0.12[b]	0.070[b]	(*S6*)
7',8'-Dihydroxuxuarine Aβ (**133**)	> 0.12[b]	> 0.12[b]	0.048[b]	(*S6*)

[a] IC_{50} (μg/ml)
[b] IC_{50} (mmol/l)

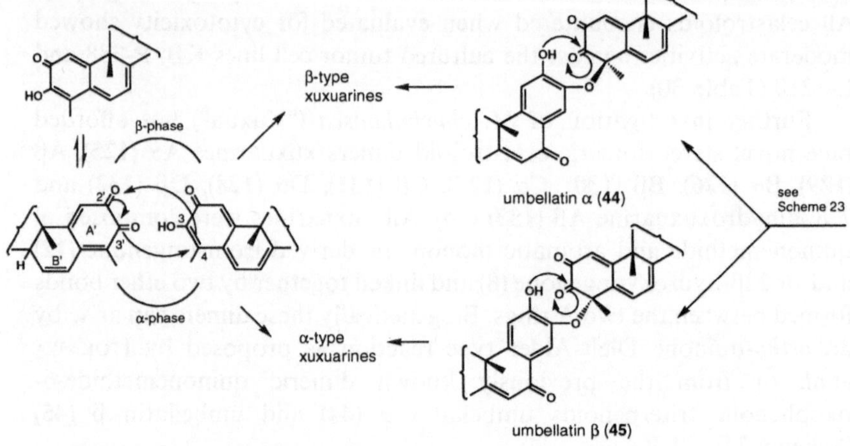

Scheme 24. Probable biogenetic origin of xuxuarines

methods. The presence of a tertiary OH function at C-3 in xuxuarine Aα **(125)** was established as a result of the appearance of an ^{1}H-NMR peak due to the amide NH of the product obtained by the treatment of **125** with trichloroacetyl isocyanate (TAI) *(S4)*. Partial assignment of the

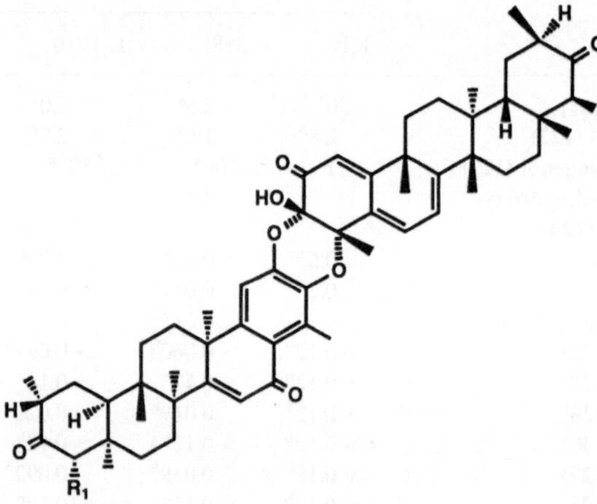

125 R$_1$ = R$_2$ = H
126 R$_1$ = R$_2$ = OH
127 R$_1$ = H, R$_2$ = OH
128 R$_1$ = OH, R$_2$ = H

^{1}H-NMR and complete assignment of ^{13}C-NMR signals of all nine xuxuarines have been made by the use of HMBC and HMQC spectra. The ^{13}C-NMR spectral assignments for selected xuxuarines are listed in Table 31. The NOESY spectrum of 3-methylxuxuarine Aα was helpful in determining the stereochemical disposition of the two monomer units in xuxuarine Aα (**125**), whereas CD spectral analysis aided in establishing the stereochemical disposition of the 3,4-dioxy linkage. Molecular dynamics (MD) and molecular mechanics (MM) calculations have been

Table 31. ^{13}C NMR Spectral Data of Selected Xuxuarines[a]

Carbon	Xuxuarine Cα (**127**) (S6)		Xuxuarine Cβ (**131**) (S6)		7',8'-Dihydroxuxuarine Aβ (**44**) (S6)	
	C	C'	C	C'	C	C'
C-1	115.60 d	111.34 d	115.05 d	110.49 d	115.14 d	108.82 d
C-2	190.26 d	144.74 s	189.45 s	145.18 s	189.45 s	145.38 s
C-3	92.26 s	137.74 s	91.17 s	137.65 s	91.19 s	137.53 s
C-4	79.43 s	127.91 s	76.90 s	128.56 s	76.85 s	129.69 s
C-5	130.38 s	124.48 s	132.15 s	123.88 s	132.31 s	125.27 s
C-6	126.64 d	187.56 s	128.61 d	187.09 s	128.39 d	200.06 s
C-7	116.34 d	126.20 d	117.28 d	126.15 d	117.19 d	37.49 t
C-8	160.18 s	170.50 s	162.95 s	170.09 s	163.23 s	41.90 d
C-9	41.60 s	39.77 s	43.52 s	39.75 s	43.65 s	37.15 s
C-10	173.72 s	150.51 s	172.78 s	151.14 s	172.71 s	152.32 s
C-11	33.48 t	34.31 t	33.32 t	34.13 t	33.16 t	33.05 t
C-12	29.80 t	30.19 t	29.85 t	30.09 t	29.88 t	29.71 t
C-13	39.40 s	40.21 s	39.75 s	40.16 s	39.84 s	39.53 s
C-14	43.98 s	44.32 s	43.67 s	44.29 s	44.00 s	44.04 s
C-15	28.09 t	28.43 t	28.22 t	28.39 t	27.92 t	28.56 t
C-16	29.48 t	35.56 t	29.47 t	35.52 t	35.52[c] t	35.38[c] t
C-17	44.84 s	38.18 s	44.78 s	38.12 s	38.22[d] s	38.16[d] s
C-18	44.95 d	43.56 d	45.00 d	43.48 d	43.54 d	44.00 d
C-19	32.08 t	31.99 t	31.91 t	31.91 t	31.96 t	31.77 t
C-20	40.80 d	41.86 d	40.83 d	41.85 d	42.31[e] d	41.95[e] d
C-21	213.48 s	213.48 s	213.44[b] s	213.37[b] s	213.91[f] s	213.32[f] s
C-22	76.47 d	52.57 t	76.35 d	52.61 t	52.41 t	53.61 t
C-23	22.22 q	12.94 q	24.55 q	13.14 q	24.61 q	13.21 q
C-25	35.63 q	38.52 q	39.78 q	38.68 q	26.39 q	39.71 q
C-26	22.37 q	20.78 q	22.36 q	20.81 q	22.39 q	14.97 q
C-27	20.83 q	19.67 q	20.39 q	19.77 q	19.64 q	18.14 q
C-28	24.97 q	32.56 q	24.94 q	32.58 q	32.77[g] q	32.52[g] q
C-30	14.70 q	15.06 q	14.70 q	15.06 q	15.17[h] q	15.11[h] q

[a]   Measured in CDCl$_3$ at 100 MHz. Chemical shifts are reported in δ units from internal standard Me$_4$Si.

[b-h] Chemical shifts bearing the same superscript in the same row are interchangeable.

carried out to confirm the orientation of the *cis* 3,4-dioxy bond of xuxuarines Aα (125) and Aβ (129) and to obtain the low energy conformations which satisfied the nOe relationship observed for these compounds. All nine xuxuarines have been subjected to cytotoxicity screening against cultured cancer cell lines. Interestingly, all were inactive in the

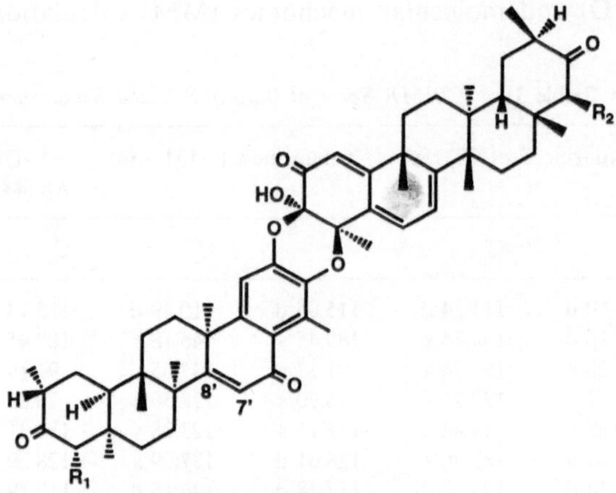

129 $R_1 = R_2 = H$

130 $R_1 = R_2 = OH$

131 $R_1 = H, R_2 = OH$

132 $R_1 = OH, R_2 = H$

133 $R_1 = R_2 = H, 7'(8')$-dihydro

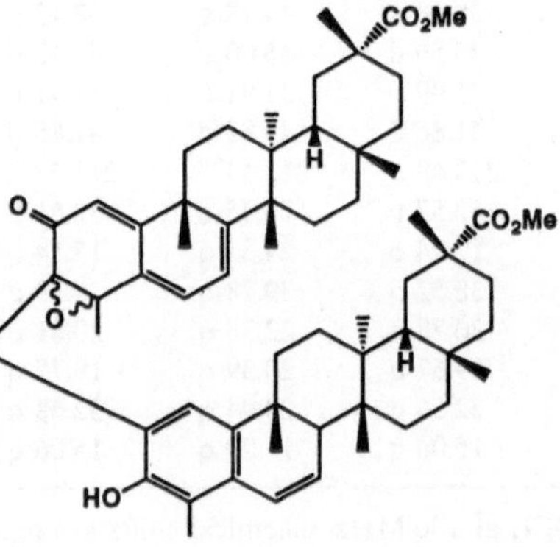

134

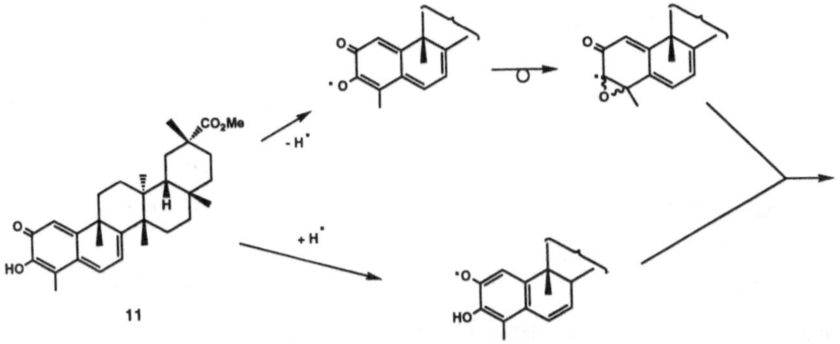

Scheme 25. Auto-oxidation of pristimerin (11) giving magellanin (134)

KB cell assay and only the α-type xuxuarines showed weak to moderate activity in P-388 and L-1210 cell lines (Table 30).

Magellanin (134) isolated from *Maytenus magellanica* (Celastraceae) belongs to a new class of dimeric quinonemethide-phenolic type of celastroloids (*S2*). Its structure elucidation involved the use of MS fragmentation, NMR (^{1}H, ^{13}C, HMBC and HMQC) and comparison of ^{1}H- and ^{13}C-NMR data with those reported for the related dimeric celastroloid cangorosin B (41; Table 7). The possibility of *in vivo* or *in vitro* production of 134 by auto-oxidation of pristimerin (11) has been suggested (Scheme 25).

STEGLICH and coworkers recently reported the isolation of a novel type of triterpenoid quinonemethide pigments russulaflavidin (135) and dihydrorussulaflavidin (136) from the hexane extract of fruiting bodies of the fungus *Russula flavida* (family: Agaricales) (*S1*). In contrast to celastroloids which bear a 24-nor-D:A-*friedo*-oleanane nucleus (Fig. 1), these quinonemethides contain a 24-nor-D:A-*friedo*-18,19-secolupane (24-nor-shionane) nucleus (Fig. 26). Russulaflavidin (135) and its dihydro derivative 136 are the first reported natural products with a 24,26-bisnorshionane skeleton.

Russulaflavidin (135) and dihydrorussulaflavidin (136) have been isolated by a combination of Sephadex LH-20 gel filtration and isocratic HPLC on silica gel under carefully defined conditions. The decolorization reaction with NaBH$_4$ together with UV, and IR data suggested some similarities between russulaflavidin (135) and pristimerin (11). Comparison of their ^{1}H- and ^{13}C-NMR data confirmed the presence of a ring A,B quinonemethide system in russulaflavidin. The total structure of 135 was determined by extensive use of NMR techniques; its absolute

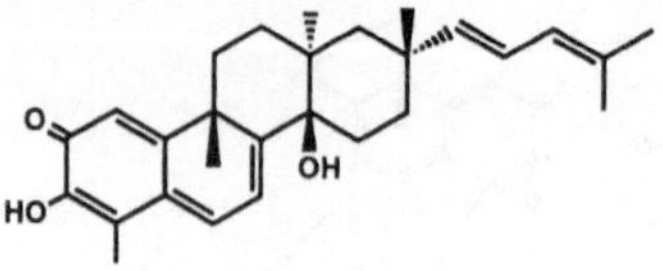

135

C$_{28}$H$_{36}$O$_3$ (420.27)

mp 78-80°C

[α]$_D$ -130° (CHCl$_3$)

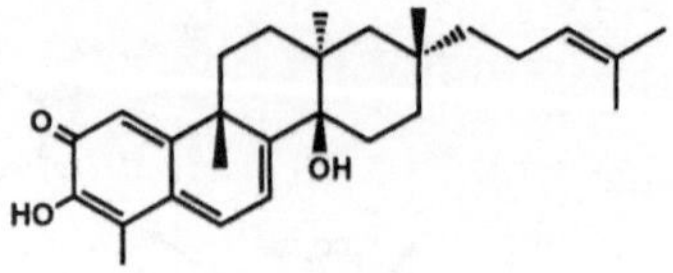

136

C$_{28}$H$_{38}$O$_3$ (422.28)

mp 81-83°C

[α]$_D$ -125° (CHCl$_3$)

Fig. 26. Physical data for russulaflavidin (135) and dihydrorussulaflavidin (136)

configuration was established by comparing its CD spectrum with that of pristimerin with the nearly identical CD spectrum (for CD spectrum of pristimerin, see Fig. 12).

Acknowledgements

The author thanks Prof. WOLFGANG STEGLICH (University of Munich, Germany) for encouraging him to write this review and Prof. DAVID G.I. KINGSTON (Virginia Polytechnic Institute and State University, USA) for encouragement and support. Research in the author's former laboratory at the Department of Chemistry, University of Peradeniya, Sri Lanka, was accomplished thanks to the skillful efforts of Dr. CHANDRA B. GAMLATH, Dr. KAMAL B. GUNAHERATH, Dr. DHAMMIKA NANAYAKKARA, Dr. GAMINI SAMARANAYAKE, Dr. W.R. WIMALASIRI, Mr. H. CHANDRASIRI FERNANDO, Ms. BHAVANI DHANABALASINGHAM, Ms. SUGANTHINI SUBRAMANIAM, and Ms. NIRMALI DIAS. Prof. TOHRU KIKUCHI (Toyama Medical and Pharmaceutical University, Japan), Dr. YASUHIRO TEZUKA (Toyama Medical and Pharmaceutical University, Japan) and Dr. VERANJA KARUNARATNE (University of Peradeniya, Sri Lanka) are thanked for collaborating with the author in his research projects on celastroloids which were funded by the University of Peradeniya (Sri Lanka), Natural Resources Science and Energy Authority (Sri Lanka), International Foundation for Science (Sweden), and International Seminar in Chemistry (Sweden). The author thanks Prof. A.G. GONZALES (University of La Laguna, Tenerife) for sending reprints of his work on celastroloids. Prof. WOLFGANG STEGLICH (University of Munich, Germany), Dr. JOHN M. RIMOLDI (Virginia Polytechnic Institute and State University, USA), Ms. MALKANTHI GUNATILAKA (Virginia Polytechnic Institute and State University, USA) are thanked for critically reading this manuscript and for helpful suggestions.

References

1. AHMED, V.U., and ATTA-UR-RAHMAN: Handbook of Natural Products Data, Vol. 2: Pentacyclic Triterpenoids. Amsterdam: Elsevier. 1994.

2. ANGELETTI, P.V., and G.B. MARINI-BETTOLO: Effect of Maitenine on RNA and Protein Synthesis in Cell Cultures. Farmaco, Ed. Sci., 29, 569 (1974); Chem. Abstr., 82, 11083y (1975).

3. BAVOVADA, R., G. BLASKO, H.-L. SHIEH, J.M. PEZZUTO, and G.A. CORDELL: Spectral Assignment and Cytotoxicity of 22-Hydroxytingenone from *Glyptopetalum sclerocarpum*. Planta Med., 56, 380 (1990).

4. BAXTER, R.L., L. CROMBIE, D.J. SIMMONDS, D.A. WHITING, O.J. BRAEDEN, and K. SZENDREI: Alkaloids of *Catha edulis* (Khat.), Part I: Isolation and Characterisation of Eleven New Alkaloids with Sesquiterpene Cores (Cathedulins): Identification of the Quinone-Methide Root Pigments. J. Chem. Soc., Perkin Trans. 1, 2965 (1979).

5. BHARGAVA, P.N: Chemical Examination of the Unsaponifiable Matter of the Fat from the Fleshy Arils of *Celastrus paniculata*. Proc. Indian Acad. Sci., 24, 506 (1946).

6. BHATNAGAR, S.S., and P.V. DIVEKAR: Pristimerin, the Anti-Bacterial Principle of *Pristimera indica*–I: Isolation, Toxicity and Anti-Bacterial Action. J. Sci. Industr. Res., 10B, 56 (1951).

7. BHATNAGAR, S.S., P.V. DIVEKAR, and N.L. DUTTA: Antibiotic Substance. Indian Patent 40,968, Feb. 28, 1951; Chem. Abstr., 45, 5884 h (1951).

8. BHATNAGAR, S.S., P.V. DIVEKAR, and N.L. DUTTA: Pristimerin. Indian Patent 40,970, Feb. 28, 1951; Chem. Abstr., 45, 5885 d (1951).

9. BROWN, P.M., M. MOIR, R.H. THOMSON, T.J. KING, V. KRISHNAMOORTHY, and T.R. SESHADRI: Tingenone and Hydroxytingenone, Triterpenoid Quinone Methides from *Euonymus tingens*. J. Chem. Soc., Perkin Trans. 1, 2721 (1973).

10. BRÜNING, R., and H. WAGNER: Übersicht über Celastraceen-Inhaltsstoffe: Chemie, Chemotaxonomie, Biosynthese, Pharmakologie. Phytochem., 17, 1821 (1978).

11. CALZADA, F., R. MATA, R. LOPEZ, E. LINARES, R. BYE, V.M. BARRETO, and F. DEL RIO: Friedelanes and Triterpenoid Quinone Methides from *Hippocratea excelsa*. Planta Med., 57, 194 (1991).

12. CAMPANELLI, A.R., M. D'ALAGNI, and G.B. MARINI-BETTOLO: Spectroscopic Evidence for the Interaction of Tingenone with DNA. FEBS Lett., 122, 256 (1980).

13. CAMPANELLI, A.R., M. D'ALAGNI, G.B. MARINI-BETTOLO, and E. GIGLIO: Optical Study of Tingenone Solubilized in Aqueous Solution of Sodium Deoxycholate. Farmaco, Ed. Pract., 36, 30 (1981); Chem. Abstr., 94, 90195 k (1981).

14. CARLISLE, C.H., and M. EHRENBERG: A Preliminary X-Ray Investigation of Pristimerin. Acta Cryst., 9, 823 (1956).

15. CHOU, T.Q., and P.F. MEI: The Principle of Chinese Drug Lei-Kung-Teng, *Tripterygium wilfordii* Hook. The Coloring Substance and the Sugars. Chinese J. Physiol., 10, 259 (1936); Chem. Abstr., 31, 1161 (1937).

16. COOKE, R.G., and R.H. THOMSON: Naturally Occurring Quinonemethines and Related Compounds. Rev. Pure Appl. Chem. (Australia), 8, 85 (1958).

17. COURTNEY, J.L., and J.S. SHANNON: Studies in Mass Spectrometry. Triterpenoids: Structure Assignment to Some Friedelane Derivatives. Tetrahedron Lett., 13 (1963).

18. DELLE MONACHE, F., G.B. MARINI-BETTOLO, O. GONCALVES DE LIMA, I.L. D'ALBUQUERQUE, and J.S. DE BARROS COELHO: Maitenin: A New Antitumoral Substance from *Maytenus* sp. Gazz. Chim. Ital., 102, 317 (1972).

19. DELLE MONACHE, F., J.F. DE MELLO, G.B. MARINI-BETTOLO, O. GONCALVES DE LIMA, and I.L. D'ALBUQUERQUE: Polpunonic Acid, A New Triterpenic Acid with Friedelane Carbon Skeleton. Gazz. Chim. Ital., 102, 636 (1972).

20. DELLE MONACHE, F., G.B. MARINI-BETTOLO, P.M. BROWN, M. MOIR, and R.H. THOMSON: The Identity of Maitenin and Tingenone. Gazz. Chim. Ital., 103, 627 (1973).

21. DELLE MONACHE, F., G.B. MARINI-BETTOLO, O. GONCALVES DE LIMA, I.L. D'ALBUQUERQUE, and J.S. DE BARROS COELHO: The Structure of Tingenone, A Quinonoid Triterpene Related to Pristimerin. J. Chem. Soc., Perkin Trans. 1, 2725 (1973).

22. DELLE MONACHE, F., M. POMPONI, G.B. MARINI-BETTOLO, and O. GONCALVES DE LIMA: Researches on Quinonoid Triterpenoids, V: 22-Hydroxytingenone. An. Quim., 70, 1040 (1974); Chem. Abstr., 83, 131782 t (1975).

23. DELLE MONACHE, F., M. POMPONI, G.B. MARINI-BETTOLO, I.L. D'ALBUQUERQUE, and O. GONCALVES DE LIMA: A Methylated Catechin and Proanthocyanidins from the Celastraceae. Phytochem., 15, 573 (1976).

24. DELLE MONACHE, F., G.B. MARINI-BETTOLO, M. POMPONI, J.F. DE MELLO, O. GONCALVES DE LIMA, and R.H. THOMSON: New Triterpene Quinone-Methides from Hippocrateaceae. J. Chem. Soc., Perkin Trans. 1, 3127 (1979).

25. DE LUCA, C., F. DELLE MONACHE, and G.B. MARINI-BETTOLO: Triterpenoid Quinones of Maytenus obtusifolia and Maytenus boaria. Rev. Latinoamer. Quim., 9, 208 (1978); Chem. Abstr., 91, 96561 m (1979).

26. DEV, S., A.S. GUPTA, and S.A. PATWARDHAN: Triterpenoids, Vol. II: Handbook of Terpenoids (SUKH DEV, ed.), pp. 552–560. Boca Raton: CRC Press. 1989.

27. DHANABALASINGHAM, B., V. KARUNARATNE, Y. TEZUKA, T. KIKUCHI, and A.A.L. GUNATILAKA: Unpublished.

28. DIAS, M.N., H.C. FERNANDO, A.A.L. GUNATILAKA, Y. TEZUKA, and T. KIKUCHI: Studies on Terpenoids and Steroids, Part 20: Isolation and NMR Analysis of 15α, 22β-Dihydroxytingenone: A Probable Biosynthetic Precursor of 14(15)-Ene-quinone-methide Triterpenoids. J. Chem. Res. (S), 238 (1990); (M), 1801 (1990).

29. DOMINGUEZ, X.A., R. FRANCO, G. CANO, S. GARCIA, and V. PENA: Plantas Medicinales Mexicanas, XXXII: Terpenoides de los Extractos Etereos de dos Celastraceas, Mortonia greggii Gray y M. palmeri Hemsl. Rev. Latinoam. Quim., 9, 33 (1978); Chem. Abstr., 89, 180174 d (1978).

30. DOMINGUEZ, X.A., R. FRANCO, G. CANO, S. GARCIA, A. ZAMUDIO, B. AMEZCUA, and X.A. DOMINGUEZ, Jr.: Triterpene Quinone-Methides from Schaefferia cuneifolia Roots. Phytochem., 18, 898 (1979).

31. DOMINGUEZ, X.A., R. FRANCO, S. GARCIA, S. AVENDANO, G. RAMIREZ, and D.A. ORTEGA: Triterpenoids from Zinowiewia integerrima, Turk (Celastraceae). Rev. Latinoam. Quim., 14, 146 (1984); Chem. Abstr., 101, 3892 g (1984).

32. DOMINGUEZ, X.A., R. FRANCO, J.B. ALCORN, S. GARCIA, Z. ORTEGA, and D. ALICIA: Medicinal Plants from Mexico. L. Pristimerin and Other Components from Crossopetalum uragoga O. Ktze (Celastraceae). Rev. Latinoam. Quim., 15, 42 (1984); Chem. Abstr., 101, 51752 h (1984).

33. EDWARDS, J.M., and A.E. SCHWARTING: Isolation of Pristimerin from Pachystima canbyi. Phytochem., 12, 945 (1973).

34. ESTRADA, R., J. CARDENAS, B. ESQUIVEL, and L. RODRIGUEZ-HAHN: D:A-Friedooleanane Triterpenes from the Roots of Acanthothamnus aphyllus. Phytochem., 36, 747 (1994).

35. FANG, S.D., D.E. BERRY, D.G. LYNN, S.M. HECHT, J. CAMPBELL, and W.S. LYNN: The Chemistry of Toxic Principles from Maytenus nemorosa. Phytochem., 23, 631 (1984).

36. FERNANDO, H.C., A.A.L. GUNATILAKA, V. KUMAR, G. WEERATUNGA, Y. TEZUKA, and T. KIKUCHI: Two New Quinone-Methides from Cassine balae; Revised Structure of Balaenonol. Tetrahedron Lett., 29, 387 (1988).

37. FERNANDO, H.C., A.A.L. GUNATILAKA, Y. TEZUKA, and T. KIKUCHI: Studies on Terpenoids and Steroids-18. Balaenonol, Balaenol and Isobalaendiol: Three New

14(15)-Ene-quinone-methide Triterpenoids from *Cassine balae*. Tetrahedron, **45**, 5867 (1989).

38. FERREIRA DE SANTANA, C., J.J. ASFORA, and C.T. CORTIAS: Primeiras Observacoes Sobre o Emprego da Maitenina em Pacientes Cancerosus. Rev. Inst. Antibioticos (Recife), **11**, 37 (1971).

39. FERREIRA DE SANTANA, C., C.T. CORTIAS, K. DE V. PINTO, M.W. SATIRO, A.L. LACERDA, and I.C. MOREIRA: Pharmacodynamic and Histophathological Studies of Maytenin. Rev. Inst. Antibioticos (Recife), **11**, 61 (1971); Chem. Abstr., **79**, 111730 p (1973).

40. FIESER, L.F., and R.N. JONES: Celastrol, Spectrographic Characterization and Color Tests. J. Amer. Pharm. Sci., Sci. Ed., **31**, 315 (1942).

41. GAMLATH, C.B., K.B. GUNAHERATH, and A.A.L. GUNATILAKA: Studies on Terpenoids and Steroids, Part 10: Structure of Four New Natural Phenolic D:A-friedo-24-Noroleanane Triterpenoids. J. Chem. Soc., Perkin Trans. 1, 2849 (1987).

42. GAMLATH, G.R.C.B., G.M.K.B. GUNAHERATH, and A.A.L. GUNATILAKA: Isolation and Structures of Some Novel Phenolic Triterpenoids of Sri Lankan Celastraceae. In: New Trends in Natural Products Chemistry (ATTA-UR-RAHMAN and P.W. LEQUESNE, eds.), p. 109. Amsterdam: Elsevier. 1986.

43. GAMLATH, C.B., and A.A.L. GUNATILAKA: Two Phenolic Friedo-23,24-Dinoroleanane Triterpenes from *Kokoona zeylanica*. Phytochem., **27**, 3221 (1988).

44. GAMLATH, C.B., A.A.L. GUNATILAKA, Y. TEZUKA, and T. KIKUCHI: The Structure of Celastranhydride: A Novel Triterpene Anhydride of Celastraceae. Tetrahedron Lett., **29**, 109 (1988).

45. GAMLATH, C.B., A.A.L. GUNATILAKA, and S. SUBRAMANIAM: Studies on Terpenoids and Steroids, Part 19: Structures of Three Novel 19(10 → 9) *abeo*-8α,9β,10α-Euphane Triterpenoids from *Reissantia indica* (Celastraceae). J. Chem. Soc., Perkin Trans. 1, 2259 (1989).

46. GAMLATH, C.B., A.A.L. GUNATILAKA, Y. TEZUKA, T. KIKUCHI, and S. BALASUBRAMANIAM: Quinone-Methide, Phenolic and Related Triterpenoids of Plants of Celastraceae: Further Evidence for the Structure of Celastranhydride. Phytochem., **29**, 3189 (1990).

47. GISVOLD, O.: The Pigments Contained in the Bark of the Root of *Celastrus scandens*, Part 1: Celastrol. J. Amer. Pharm. Assoc., **28**, 440 (1939).

48. GISVOLD, O.: The Constitution of Celastrol, Part II. J. Amer. Pharm. Assoc., **29**, 12 (1940).

49. GISVOLD, O.: The Constitution of Celastrol, Part III. J. Amer. Pharm. Assoc., **29**, 432 (1940).

50. GISVOLD, O.: The Constitution of Celastrol, Part IV. J. Amer. Pharm. Assoc., **31**, 529 (1942).

51. GONCALVES DE LIMA, O., I.L. D'ALBUQUERQUE, J.S. DE BARROS COELHO, G.M. MACIEL, M. DA S.B. CALVALCANTI, D.G. MARTINS, and A.L. LACERDA: Antimicrobial Substances from Higher Plants, XXX: Antimicrobial and Antineoplastic Activity of Pristimerin Isolated from *Prionostemma aspera*, from the Humid Bushes of the Pernambuco Region. Rev. Inst. Antibioticos (Recife), **9**, 3 (1969); Chem. Abstr., **75**, 95369 n (1971).

52. GONCALVES DE LIMA, O., I.L. D'ALBUQUERQUE, J.S. DE BARROS COELHO, D.G. MARTINS, A.L. LACERDA, and G.M. MACIEL: Antimicrobial Substances from Higher Plants, XXXI: Antimicrobial and Antitumoral Activity of Maytenin Isolated from the Cortical Root Zone of *Maytenus* Species. Rev. Inst. Antibioticos (Recife), **9**, 17 (1969); Chem. Abstr., **75**, 95370 f (1971).

53. GONCALVES DE LIMA, O., I.L. D'ALBUQUERQUE, and G.M. MACIEL: Antimicrobial Substances in Higher Plants, XXIX: Antibiotic Activity of Celastrol. Rev. Inst. Antibioticos (Recife), **9**, 75 (1969); Chem. Abstr., **75**, 106532 p (1971).

54. GONCALVES DE LIMA, O., J.S. DE BARROS COELHO, E. WEIGERT, I.L. D'ALBUQUERQUE, D. DE A. LIMA, and M.A. DE M. SOUZA: Antimicrobial Compounds from Higher Plants, XXXVI: Maytenin and Pristimerin in the Root Cortex of *Maytenus ilicifolia* from Southern Brazil. Rev. Inst. Antibioticos (Recife), **11**, 35 (1971); Chem. Abstr., **77**, 29630 f (1972).

55. GONCALVES DE LIMA, O., E. WEIGERT, G.B. MARINI-BETTOLO, G.M. MACIEL, and J.S. DE BARROS COELHO: Antimicrobial Substances of Higher Plants, XXXVIII: Isolation and Identification of Maitenin and Pristimerin from Roots of *Plenckia populnea* Collected from Woods Near University of Brasilia. Rev. Inst. Antibioticos (Recife), **12**, 19 (1972); Chem. Abstr., **81**, 148442 p (1974).

56. GONCALVES DE LIMA, O., J.S. DE BARROS COELHO, G.M. MACIEL, E.P. HERINGER, and C. GONCALVES DE LIMA: Antimicrobial Substances of Higher Plants, XXXIX: Identification of Pristimerin as an Active Component of Bacupari (*Salacia crassifolia*) from the Araguaya Region of Brazil. Rev. Inst. Antibioticos (Recife), **12**, 19 (1972); Chem. Abstr., **81**, 148443 q (1974).

57. GONZALEZ, A., C.G. FRANCISCO, R. FREIRE, R. HERNANDEZ, J.A. SALAZAR, and E. SUAREZ: Iguesterin, A New Quinoid Triterpene from *Catha cassinoides*. Phytochem., **14**, 1067 (1975).

58. GONZALEZ, A.G., V. DARIAS, J. BOADA, and G. ALONSO: Study of the Cytotoxic Activity of Iguesterin and Related Compounds. Planta Med., **32**, 282 (1977).

59. GONZALEZ, A.G., B.M. FRAGA, P. GONZALEZ, C.M. GONZALEZ, A.G. RAVELO, E. FERRO, X.A. DOMINGUEZ, M. MARTINEZ, A. PERALES, and J. FAYOS: Crystal Structure of Orthosphenic Acid. J. Org. Chem., **48**, 3759 (1983).

60. GONZALEZ, A.G., G.D. MONACHE, F.D. MONACHE, and G.B. MARINI-BETTOLO: Chuchuhuasa – A Drug Used in Folk Medicine in the Amazonian and Andean Areas. A Chemical Study of *Maytenus laevis*. J. Ethnopharmacol. **5**, 73 (1982).

61. GONZALEZ, A.G., B.M. FRAGA, C.M. GONZALEZ, A.G. RAVELO, E. FERRO, X.A. DOMINGUEZ, M.A. MARTINEZ, J. FAYOS, A. PERALES, and M.L. RODRIGUEZ: X-Ray Analysis of Netzahualcoyone, A Triterpene Quinone Methide from *Orthosphenia mexicana*. Tetrahedron Lett., **24**, 3033 (1983).

62. GONZALEZ, A.G., I. LOPEZ, E.A. FERRO, A.G. RAVELO, J. GUTIERREZ, and M.A. AGUILAR: Taxonomy and Chemotaxonomy of Some Species of Celastraceae. Biochem. Syst. Ecol., **14**, 479 (1986).

63. GONZALES, A.G., A.G. RAVELO, B.E. FERRO, C.M. GONZALEZ, J.G. LUIS, I.L. BAZZOCCHI, and R.L. DORTA: Plantas Iberamericanas como Fuentes de Moleculas Bioactivas, I: Celastraceas. Santa Cruz de Tenerife: Fundacion AIETI, OEI, Litigrafia A. Romero. 1987.

64. GONZALEZ, A.G., I.L. BAZZOCCHI, A.G. RAVELO, J.G. LUIS, X.A. DOMINGUEZ, G. VAZQUEZ, and G. CANO: Triterpenes and Triterpenoquinones of *Rzedowskia tolantoguensis*. Rev. Latinoam. Quim., **18**, 83 (1987); Chem. Abstr., **108**, 147100 q (1988).

65. GONZALEZ, A.G., C.M. GONZALEZ, E.A. FERRO, A.G. RAVELO, and X.A. DOMINGUEZ: Triterpenes from Celastraceae. J. Chem. Res. (S) 20, (M) 273 (1988).

66. GONZALEZ, A.G., A.G. RAVELO, I.L. BAZZOCCHI, J. JIMENEZ, C.M. GONZALEZ, J.G. LUIS, E.A. FERRO, A. GUTIERREZ NAVARRO, L. MOUJIR, and F.G. DE LAS HERAS: Biological Study of Triterpene Quinones from Celastraceae. Farmaco, Ed. Sci., **43**, 501 (1988); Chem. Abstr., **110**, 264 d (1988).

67. GONZALEZ, A.G., C.M. GONZALEZ, A.G. RAVELO, A.N. GUTIERREZ, L. MOUJIR, E.

NAVARRO, and J. BOADA: Netzahualcoyone, A Triterpenequinone, Biological Activity. Rev. Latinoam. Quim., **19**, 36 (1988); Chem. Abstr., **109**, 104764 n (1988).

68. GONZALEZ, A.G., R.L. DORTA, A.G. RAVELO, and J.G. LUIS: Synthesis of Friedelane Triterpenes with a 24-Hydroxy-3-oxo-hemiacetal Group. J. Chem. Res. (S), 150 (1988), (M), 1228 (1988).
69. GONZALEZ, A.G., J.J. MENDOZA, J.G. LUIS, A.G. RAVELO, and L.L. BAZZOCCHI: New Epimeric Di-Triterpene Quinone Ethers. Their Partial Synthesis and That of Netzahualcoyene from Pristimerin and DDQ. Tetrahedron Lett., **30**, 863 (1989).
70. GONZALEZ, A.G., and A. GUTIERREZ: Phytochemistry of Celastraceae. Rev. R. Acad. Cienc. Exactas, Fis. Nat. Madrid, **84**, 237 (1990); Chem. Abstr., **117**, 44489e (1992).
71. GONZALEZ, A.G., J.J. MENDOZA, A.G. RAVELO, J.G. LUIS, and X.A. DOMINGUEZ: New Pentacyclic Triterpenes of *Schaefferia cuneifolia* (Celastraceae). Rev. Latinoam. Quim., **22**, 3 (1991); Chem. Abstr., **116**, 55559 h (1992).
72. GONZALEZ, A.G., J.S. JIMENEZ, L.M. MOUJIR, A.G. RAVELO, J.G. LUIS, L.L. BAZZOCCHI, and A.M. GUTIERREZ: Two New Triterpene Dimers from Celastraceae, Their Partial Synthesis and Antimicrobial Activity. Tetrahedron, **48**, 769 (1992).
73. GRANT, P.K., and A.W. JOHNSON: Pristimerin, Part I: The Nature of the Chromophore. J. Chem. Soc., 4079 (1957).
74. GRANT, P.K., and A.W. JOHNSON: Pristimerin, Part II: Further Reactions Involving the Chromophore. J. Chem. Soc., 4669 (1957).
75. GRANT, P.K., A.W. JOHNSON, P.F. JUBY, and T.J. KING: Pristimerin, Part III: A Modified Structure for the Chromophore. J. Chem. Soc., 549 (1960).
76. GUNAHERATH, G.M.K.B., A.A.L. GUNATILAKA, M.U.S. SULTANBAWA, and M.I.M. WAZEER: The Structure of Zeylasterone – The First of a Series of Phenolic 24-Nor-D: A-friedo-oleanane Triterpenes. Tetrahedron Lett., **21**, 4749 (1980).
77. GUNAHERATH, G.M.K.B., and A.A.L. GUNATILAKA: Structures of Two New Phenolic 24-Nor-D: A-friedo-oleananes Related to Zeylasterone: A Partial Synthesis of Trimethylzeylasterone. Tetrahedron Lett., **24**, 2025 (1983).
78. GUNAHERATH, G.M.K.B., and A.A.L. GUNATILAKA: Studies on Terpenoids and Steroids, Part 3: Structure and Synthesis of a New Phenolic D: A-Friedo-24-noroleanane Triterpenoid, Zeylasterone, from *Kokoona zeylanica*. J. Chem. Soc., Perkin Trans. 1, 2845 (1983).
79. GUNAHERATH, G.M.K.B., and A.A.L. GUNATILAKA: 23-Oxo-isopristimerin III: A New Natural Phenolic (9 → 8)-24-Nor-D: A-friedo-oleanan Triterpene. Tetrahedron Lett., **24**, 2799 (1983).
80. GUNATILAKA, A.A.L.: Triterpenoids and Steroids of Sri Lankan Plants: A Review of Occurrence and Chemistry. J. Nat. Sci. Coun. Sri Lanka, **14**, 1 (1986); Chem. Abstr., **109**, 20229 d (1988).
81. GUNATILAKA, A.A.L., C. FERNANDO, T. KIKUCHI, and Y. TEZUKA: ^{1}H and ^{13}C NMR Analysis of Three Quinone-Methide Triterpenoids. Magn. Reson. Chem., **27**, 803 (1989).
82. GUNATILAKA, A.A.L., and W.R. WIMALASIRI: Studies on Terpenoids and Steroids, Part 22: Structure and Some Reactions of Pristimerin Leucotriacetate. J. Chem. Res. (S), 30 (1992).
83. GUNATILAKA, A.A.L., and N.P.D. NANAYAKKARA: Studies on Terpenoids and Steroids, 2: Structures of Two New Tri- and Tetra-Oxygenated D: A-Friedo-oleanan Triterpenes from *Kokoona zeylanica*. Tetrahedron, **40**, 805 (1984).
84. GUNATILAKA, A.A.L.: Recent Studies on Some Medicinal and Related Plants of Sri Lanka. In: Studies in Natural Products Chemistry, Vol. 5: Structure Elucidation, Part B (ATTA-UR-RAHMAN, ed.), p. 743. Amsterdam: Elsevier. 1989.

85. GUNATILAKA, A.A.L., K.B. GUNAHERATH, S. SOTHEESWARAN, S. BALASUBRAMANIAM, V. GUNAWARDENA, and K. JAYASENA: Unpublished.
86. HALLER, H.L., E.H. SIEGLER, and M.C. SWINGLE: A Chinese Insecticidal Plant, *Tripterygium wilfordii*, Introduced into the United States. Science, **93**, 60 (1941).
87. HAM, P.J., and D.A. WHITING: X-Ray Analysis of Pristimerol Bis-*p*-bromobenzoate, A Derivative of the Triterpene Quinone Methide Pristimerin. J. Chem. Soc., Perkin Trans. 1, 330 (1972).
88. HARADA, R., H. KAKISAWA, S. KOBAYASHI, M. MUSYA, K. NAKANISHI, and Y. TAKAHASHI: Structure of Pristimerin, A Quinonoid Triterpene. Tetrahedron Lett., 603 (1962).
89. HEGNAUER, R.: Chemotaxonomie der Pflanzen, Band 4, p. 262. Basel: Birkhäuser. 1966.
90. HEYWOOD, V.H.: Flowering Plants of the World, p. 178. New York: Oxford University Press. 1993.
91. IRELAND, R.E., and D.M. WALBA: The Total Synthesis of (−)-Friedelin, An Unsymmetrical, Pentacyclic Triterpene. Tetrahedron Lett., 1071 (1976).
92. ITOKAWA, H., O. SHIROTA, H. MORITA, K. TAKEYA, N. TOMIOKA, and A. ITAI: New Triterpene Dimers from *Maytenus ilicifolia*. Tetrahedron Lett., **31**, 6881 (1990).
93. ITOKAWA, H., O. SHIROTA, H. IKUTA, H. MORITA, K. TAKEYA, and Y. IITAKA: Triterpenes from *Maytenus ilicifolia*. Phytochem., **30**, 3713 (1991).
94. JAIN, M.K.: Chemical Examination of *Celastrus paniculatus* Willd. Indian J. Chem., **1**, 500 (1963).
95. JOHNSON, A.W., P.F. JUBY, T.J. KING, and S.W. TAM: Pristimerin, Part IV: Total Structure. J. Chem. Soc., 2884 (1963).
96. JOSHI, K.C., R.K. BANSAL, and R. PATNI: Chemical Constituents of *Gymnosporia montana* and *Euonymus pendulus*. Planta Med., **34**, 211 (1978).
97. JOSHI, K.C., P. SINGH, and C.L. SINGHI: Triterpenoid Quinone-Methides from *Gymnosporia montana*. Planta Med., **43**, 89 (1981).
98. KAMAT, V.N., F. FERNANDES, and S.S. BHATNAGAR: Chemical Constitution and Biological Activity of Pristimerin. J. Sci. Industrial Res., **14C**, 1 (1955).
99. KRISHNAMOORTHY, V., J.D. RAMANATHAN, and T.R. SESHADRI: Constitution of Tingenone, A Component of the Stembark of *Euonymus tingens* Wall. (F. Celastraceae). Tetrahedron Lett., 1047 (1962).
100. KRISHNAMOORTHY, V., T.R. SESHADRI, R.H. THOMSON, and M. MOIR: IUPAC 8th International Symposium on the Chemistry of Natural Products, New Delhi. Abstracts of Papers, p. 165. 1972.
101. KRISHNAN, V., and S. RANGASWAMI: Quinone Components of *Salacia macrosperma*: Constitution of Saptarangi Quinone-A. Indian J. Chem., **9**, 117 (1971).
102. KULKARNI, A.B., and R.C. SHAH: Structure of Pristimerin. Nature (London), **173**, 1237 (1954).
103. KUMAR, V., M.I.M. WAZEER, and D.B.T. WIJERATNE: 21α,26-Dihydroxy-D:A-friedooleanan-3-one from *Salacia reticulata* var. *diandra* (Celastraceae). Phytochem., **24**, 2067 (1985).
104. KUTNEY, J.P., M.H. BEALE, P.J. SALISBURY, K.L. STUART, B.R. WORTH, P.M. TOWNSLEY, W.T. CHALMERS, K. NILSSON, and G.G. JACOLI: Isolation and Characterization of Natural Products from Plant Tissue Cultures of *Maytenus buchananii*. Phytochem., **20**, 653 (1981).
105. KUTNEY, J.P., G.M. HEWITT, T. KURIHARA, P.J. SALISBURY, R.D. SINDELAR, K.L. STUART, P.M. TOWNSLEY, W.T. CHALMERS, and G.G. JACOLI: Cytotoxic Diterpenes from Tissue Cultures of *Tripterygium wilfordii*. Can. J. Chem., **59**, 2677 (1981).

106. LEWIS, R.J.: Sax's Dangerous Properties of Industrial Materials, 8th Ed. PMO 525. London: Van Nostrand-Reinhold. 1992.

107. LIKHITWITAYAWUID, K., R. BAVOVADA, L.-Z. LIN, and G.A. CORDELL: Revised Structure of 20-Hydroxytingenone and ^{13}C NMR Assignments of 22β-Hydroxytingenone. Phytochem., **34**, 759 (1993).

108. MAHATO, S.B., A.K. NANDY, and G. ROY: Triterpenoids. Phytochem., **31**, 2199 (1992).

109. MARINI-BETTOLO, G.B.: Chemistry of the Active Principles of Celastraceae-Hippocrataceae. Farmaco Ed. Sci., **29**, 551 (1974); Chem. Abstr., **82**, 13925 y (1975).

110. MARINI-BETTOLO, G.B.: A Particular Group of Natural Substances. Triterpene Phenoldienones. Rev. Latinoamer. Quim., **10**, 97 (1979); Chem. Abstr., **92**, 18720 u (1980).

111. MARTA, M., F. DELLE MONACHE, G.B. MARINI-BETTOLO, J.F. DE MELLO, and O. GONCALVES DE LIMA: Rigidenol. A New Triterpene with a Lupane Skeleton from *Maytenus rigida*. Gazz. Chim. Ital., **109**, 61 (1979).

112. MARTIN, J.D.: The Structure of Dispermoquinone. A Triterpenoid Quinone Methide from *Maytenus dispermus*. Tetrahedron, **29**, 2997 (1973).

113. MARTINOD, P., A. PAREDES, F. DELLE MONACHE, and G.B. MARINI-BETTOLO: Isolation of Tingenone and Pristimerin from *Maytenus chuchuhuasca*. Phytochem., **15**, 562 (1976).

114. MELO, A.M., M.L. JARDIM, C.F. DE SANTANA, Y. LACET, J.L. FILHO, O. GONCALVES DE LIMA, and I.L. D'ALBUQUERQUE: Primeras Observacoes do Uso Topico da Primina, Plumbagina e Maitenina em Pacientes Portadores de Cancer de Pele. Rev. Inst. Antibiotics (Recife), **14**, 9 (1974).

115. MOUJIR, L., A.M. GUTIERREZ-NAVARRO, A.G. GONZALEZ, A.G. RAVELO, and J.G. LUIS: The Relationship Between Structure and Antimicrobial Activity in Quinones from the Celastraceae. Biochem. Syst. Ecol., **18**, 25 (1990).

116. MOUJIR, L., A.M. GUTIERREZ-NAVARRO, A.G. GONZALEZ, A.G. RAVELO, and J.G. LUIS: Mode of Action of Netzahualcoyone. Antimicrobial Agents and Chemotherapy, **35**, 211 (1991).

117. NAKANISHI, K., H. KAKISAWA, and Y. HIRATA: Structure of Pristimerin and Celastrol. J. Amer. Chem. Soc., **77**, 3169 (1955); Erratum: J. Amer. Chem. Soc., **77**, 6729 (1955).

118. NAKANISHI, K., H. KAKISAWA, and Y. HIRATA: Structure of Pristimerin and Celastrol. Bull. Chem. Soc. Japan, **29**, 7 (1956).

119. NAKANISHI, K., Y. TAKAHASHI, and H. BUDZIKIEWICZ: Pristimerin. Spectroscopic Properties of the Dienone-Phenol-Type Rearrangement Products and Other Derivatives. J. Org. Chem., **30**, 1729 (1965).

120. NAKANISHI, K., V.P. GULLO, L. MIURA, T.R. GOVINDACHARI, and N. VISWANATHAN: Structure of Two Triterpenes. Application of Partially Relaxed Fourier Transform ^{13}C Nuclear Magnetic Resonance. J. Am. Chem. Soc., **95**, 6493 (1973).

121. NAKANISHI, K.: A Wandering Natural Products Chemist – Profiles, Pathways and Dreams (J.I. SEEMAN, ed.), p. 34. Washington, DC: American Chemical Society. 1991.

122. NGASSAPA, O., D.D. SOEJARTO, J.M. PEZZUTO, and N.R. FARNSWORTH: Quinone-Methide Triterpenes and Salaspermic Acid from *Kokoona ochracea*. J. Nat. Prod., **57**, 1 (1994).

123. PANT, P., and R.P. RASTOGI: The Triterpenoids. Phytochem., **18**, 1095 (1979).

124. PATRA, A., and S.K. CHAUDHURI: Assignment of Carbon-13 Nuclear Magnetic Resonance Spectra of Some Friedelanes. Magn. Reson. Chem., **25**, 95 (1987).

125. PAVANAND, K., H.K. WEBSTER, K. YONGVANITCHIT, A. KUN-ANAKE, T. DECHATIWONGSE, W. NUTAKUL, and J. BANSIDDHI: Schizontocidal Activity of *Celastrus paniculatus* Willd. Against *Plasmodium falciparum in vitro*. Phytother. Res., **3**, 136 (1989).

126. POMPONI, M., F. DELLE MONACHE, and G.B. MARINI-BETTOLO: Rearrangements of Tingenone IV. Researches on Quinonoid Triterpenes. Anales de Quimica, 1037 (1974); Chem. Abstr., **83**, 131781 s (1975).

127. PREMAKUMARA, G.A.S., W.D. RATNASOORIYA, S. BALASUBRAMANIAM, B. DHANABALASINGHAM, H.C. FERNANDO, M.N. DIAS, V. KARUNARATNE, and A.A.L. GUNATILAKA: Studies on Terpenoids and Steroids, Part 24: The Effect of Some Natural Quinonemethide and 14(15)-Ene-quinonemethide Nortriterpenoids on Motility of Human Spermatozoa *in vitro*. Phytochem. (Life Sci. Adv.), **11**, 219 (1992).

128. RAVELO, A.G., J.G. LUIS, C.M. GONZALEZ, E.A. FERRO., I.L. BAZZOCCHI, J. JIMENEZ, H.R. HERRERA, A. JIMENEZ, and Z.E. AGUIAR: Chemical-Taxonomic Relations Between Lamiaceae and Celastraceae. Rev. Latinoam. Quim., **19**, 72 (1988); Chem. Abstr., **110**, 54443 e (1988).

129. REDDY, G.C.S., K.N.N. AYENGAR, and S. RANGASWAMI: Salacia Quinonemethide: A New Compound Related to Pristimerin from the Root Bark of *Salacia macrosperma* Wight. Indian J. Chem., **14B**, 131 (1976).

130. REDDY, G.C.S., K.N.N. AYENGAR, and S. RANGASWAMI: Chemical Components of *Salacia macrosperma* Wight and Structure of Salacia Quinonemethide. Ind. J. Chem., **20B**, 197 (1981).

131. ROBSON, N.: Taxonomic and Nomenclatural Notes on Celastraceae. Bot. Soc. Brot., Ser. 2, **39**, 5 (1965).

132. SCHECHTER, M.S., and H.L. HALLER: Identity of the Red Pigment in the Roots of *Tripterygium wilfordii* and *Celastrus scandens*. J. Am. Chem. Soc., **64**, 182 (1942).

133. SCHWENK, E.: Tumor Action of Some Quinonoid Compounds in the Cheekpouch Test. Arzneim.-Forsch., **12**, 1143 (1962).

134. SESHADRI, S., V.V. MHASKAR, A.B. KULKARNI, and R.C. SHAH: Pristimerin, Part II. J. Sci. Industr. Res., **17B**, 111 (1958).

135. SHAH, R.C., A.B. KULKARNI, and V.M. THAKORE. Pristimerin, Part I. J. Chem. Soc., 2515 (1955).

136. SNEDEN, A.T.: Isoiguesterin, A New Antileukemic Bisnortriterpene from *Salacia madagascariensis*. J. Nat. Prod., **44**, 503 (1981).

137. DE SOUZA, J.R., N.K. JANNOTTI, G.D.F. SILVA, and J.A. PINHEIRO: Friedelane Triterpene Related to Populnonic Acid. Gazz. Chim. Ital., **118**, 821 (1988).

138. DE SOUZA, J.R., G.D.F. SILVA, J.L. PEDERSOLI, and R.J. ALVES: Friedelane and Oleanane Triterpenoids from Bark Wood of *Austroplenckia populnea*. Phytochem., **29**, 3259 (1990).

139. TEZUKA, Y., T. KIKUCHI, C.B. GAMLATH, and A.A.L. GUNATILAKA: Studies on Terpenoids and Steriods, Part 17: ^{1}H and ^{13}C NMR Studies of Two Natural Phenolic D:A-Friedo-24-Noroleanane Triterpenoids. J. Chem. Res. (S), 268 (1989), (M), 1901 (1989).

140. TEZUKA, Y., T. KIKUCHI, H.C. FERNANDO, and A.A.L. GUNATILAKA: ^{1}H and ^{13}C NMR Spectral Assignments of Some Ene-quinonemethide Nortriterpenoids. Phytochem., **32**, 1531 (1993).

141. TEZUKA, Y., T. KIKUCHI, B. DHANABALASINGHAM, V. KARUNARATNE, and A.A.L. GUNATILAKA: Salacenonal: A Novel Nortriterpenoid Aldehyde of Biogenetic Significance from *Salacia reticulata*. Nat. Prod. Lett., **3**, 273 (1993).

142. TEZUKA, Y., T. KIKUCHI, B. DHANABALASINGHAM, V. KARUNARATNE, and A.A.L. GUNATILAKA: Studies on Terpenoids and Steroids, 25: Complete ^{1}H and ^{13}C NMR Spectral Assignments of Salaciquinone, A New 7-Oxo-quinonemethide Dinortriterpenoid. J. Nat. Prod., **57**, 270 (1994).

143. TURNER, A.B.: Quinone Methides in Nature. In: Progress in the Chemistry of Organic Natural Products (L. ZECHMEISTER, ed.), **24**, 288 (1966).
144. TURNER, A.B.: Quinone Methides. Quart. Rev. (Chem. Soc. London), **18**, 347 (1964).
145. VISWANATHAN, N.I.: Salaspermic Acid, A New Triterpene Acid from *Salacia macrosperma* Wight. J. Chem. Soc., Perkin Trans. 1, 349 (1979).
146. WATANABE, T., A. MIYAKE, M. GOTO, and S. HORII: Antibacterial and Antitumor Substance from *Celastrus articulatus*. Japanese Patent 18,393 ('60), Dec. 19, 1961; Chem. Abstr., **55**, 20339 d (1962).
147. WIJERATNE, D.B.T., V. KUMAR, M.U.S. SULTANBAWA, and S. BALASUBRAMANIAM: Triterpenoids from *Gymnosporia emarginata*. Phytochem., **21**, 2422 (1982).
148. WOODLAND, D.W.: Contemporary Plant Systematics, p. 264. New Jersey: Prentice-Hall. 1991.

Addendum References

S1. FRÖDE, R., M. BRÖCKELMANN, B. STEFFAN, W. STEGLICH, and R. MARUMOTO: A Novel Type of Triterpenoid Quinone Methide Pigment from the Toadstool *Russula flavida* (Agaricales). Tetrahedron, **51**, 2553 (1995).
S2. GONZALEZ, A.G., A. CRESPO, A.G. RAVELO, and O.M. MUNOZ: Magellanin, A New Triterpene Dimer from *Maytenus magellanica* (Celastraceae). Nat. Prod. Lett., **4**, 165 (1994).
S3. GONZALEZ, A.G., N.L. ALVARENGA, A.G. RAVELO, I.A. JIMENEZ, and I.L. BAZZOCCHI: Two Triterpenes from *Maytenus canariensis*. J. Nat. Prod., **58**, 570 (1995).
S4. GOODLETT, V.W.: Use of In Situ Reactions for Characterization of Alcohols and Glycols by Nuclear Magnetic Resonance. Anal. Chem., **37**, 431 (1965).
S5. SHIROTA, O., H. MORITA, K. TAKEYA, H. ITOKAWA, and Y. IITAKA: Cytotoxic Aromatic Triterpenes from *Maytenus ilicifolia* and *Maytenus chuchuhuasca*. J. Nat. Prod., **57**, 1675 (1994).
S6. SHIROTA, O., H. MORITA, K. TAKEYA, and H. ITOKAWA: Structures of Xuxuarines, Stereoisomeric Triterpene Dimers from *Maytenus chuchuhuasca*. Tetrahedron, **51**, 1107 (1995).

(Received January 10, 1995)

The Spirostaphylotrichins and Related Microbial Metabolites

P. Walser-Volken and Ch. Tamm,
Institut für Organische Chemie, Universität Basel, Switzerland

Contents

1. Introduction. 125
2. Isolation of Spirostaphylotrichin A (1), B (2), C and D (3/4), Q (5), R (6) and
   F (7) from Cultures of *Staphylotrichum coccosporum* 127
3. Biosynthetic Studies . . . . : . . . 135
   3.1. Feeding Experiments with ^{14}C-, ^{13}C, and ^{2}H-Labelled Precursors 135
       3.1.1. Experiments with ^{14}C-Labelled Precursors 135
       3.1.2. Experiments with ^{13}C- and ^{2}H-Labelled Precursors 136
   3.2. Investigation of Mutant Strains and Isolation of Further
       Spirostaphylotrichins. 142
       3.2.1. Mutagen Treatment and Selection of Mutants. 142
       3.2.2. The Mutant Strain *P 84*. 143
       3.2.3. The Mutant Strain *P 649* 146
4. Investigation of the Epimerization of Spirostaphylotrichins 150
5. Discussion of the Results Regarding the Biosynthesis of the
   Spirostaphylotrichins. 153
6. Synthetic Approaches Towards the Spirostaphylotrichins 156
7. Related Fungal Metabolites: Triticones, Arthropsolides, and Other Compounds 159
   7.1. Triticones. 159
   7.2. Arthropsolides and Related Compounds 160
References. 164

1. Introduction

Fungi are organisms widely distributed in nature and particularly abundant in soils, damp environments and in bad rests of foodstuffs. They encompass the most various forms of life (unicellular, multicellular, spherical, filamentous, parasitic, saprobic, symbiotic, *etc.*) and are often considered to be harmful, pathogenic or even toxic. Nevertheless fungi possess, besides their ecological importance as destruents (in connection with the decomposition of organic materials), also a great utility as producers of pharmacologically and commercially valuable natural substances. Indeed intensive research over the past five decades has shown that soil derived fungi, and above all those assigned to the class of *Deuteromycetes* (*Fungi imperfecti*), represent a rich source of bioactive secondary metabolites.

The tropical fungus *Staphylotrichum coccosporum*, discovered in 1957 in soil samples taken from regions of Zaire and Mexico, belongs to the *Deuteromycetes* too, more exactly to the order *Hyphomycetes* (*1*). In the beginning of the eighties PETER and AUDEN (*2*) isolated from submerged cultures of *S. coccosporum* a new secondary metabolite exhibiting a remarkable lipid-lowering activity. The structure of this unknown natural product, later called spirostaphylotrichin A (**1**), was assigned on the basis of spectroscopic data (see Table 1). Its absolute configuration was established by means of CD-spectroscopic measurements of its 4,6-O-bis-4'-methoxybenzoyl derivative (*3*).

The novel structure of this spirocyclic substance, with its unusual substituted γ-lactam moiety, promped SANDMEIER and TAMM to investigate its biosynthesis. To this end, three methods were used: (1) The production of secondary metabolites of cultures of *S. coccosporum* in different nutrient media was examined in order to eventually identify and isolate other substances biogenetically related to spirostaphylotrichin A (**1**), (2) A series of incorporation experiments was carried out, using radioactive as well as ^{13}C- and ^{2}H-labelled potential precursors, (3) Mutant strains of *S. coccosporum* blocked in the biosynthesis of **1** were produced by UV irradiation, cultivated under optimal conditions and tested with regard to the accumulation of possible intermediates in the biogenetic pathway.

Some results of these studies have been published in preliminary communications (*3–6*). In this review article a complete account of the investigations in this field is provided. In addition a proposal for a pos-

A (**1**)

sible biosynthetic pathway for spirostaphylotrichin A (1) is presented in the light of new experimental findings. In a subsequent chapter synthetic approaches are summarized and finally some other fungal metabolites are discussed, which have recently been isolated by SUGAWARA (7, 8) and AYER (9, 10) respectively, and that exhibit interesting structural and probably also biogenetic similarities with the spirostaphylotrichins.

2. Isolation of Spirostaphylotrichin A (1), B (2), C and D (3/4)[1], Q (5), R (6) and F (7) from Cultures of *Staphylotrichum coccosporum*

Cultures of the wild type strain *S. coccosporum DMS 2602*[2] were grown according to a multi-stage fermentation method already described by SANDMEIER and TAMM in a precedent paper (3). This cultivation procedure is briefly summarized. The fungus was first inoculated on 25 ml agar medium in a 100-ml Erlenmeyer flask and incubated at 28°, in the dark, during 5–7 days. Then 25 ml of N1 153 medium (3) were added to the mycelium previously separated from the agar. This preculture was incubated for further 24 hours on a rotary shaker (200 rpm) and was utilized afterwards to inoculate a 500-ml Erlenmeyer flask containing 180 ml of N1 153 medium. After approximately 48 hours of fermentation at 28° and under continuous shaking, portions of this second preculture were used for the inoculation of production cultures[3]. With the intention to investigate the metabolite production of *S. coccosporum* under different nutritional conditions, production cultures were grown in two kinds of media: a complex one, on the basis of equal amounts of soya meal and D-mannitol, and a minimal one, containing D-glucose, NH_4NO_3, $MgHPO_4$, amino acids, vitamins and some trace elements.

The course of the accumulation of spirostaphylotrichin A (1) and of new compounds in the culture broths was followed at regular intervals by HPLC-analysis of crude extracts in CH_2Cl_2 obtained from aliquots of

---

[1] The structures of spirostaphylotrichin C and D (3/4) may be interchanged.

[2] This strain is registered under the number 2602 at the German collection of microorganisms (Deutsche Sammlung von Mikroorganismen) in Göttingen. We thank Dr. H. H. PETER and J. A. L. AUDEN, Ciba-Geigy AG, Basel, Switzerland, for providing the strain and advice for growing cultures.

[3] Ca. 12–15 g of the second preculture were used for the inoculation of single shaking production cultures. 700 ml of this preculture are necessary for the inoculation of a 11-litre fermenter.

the respective cultures. This analysis of the metabolites was carried out on a *reversed phase* column (0.4 × 25 cm, Spherisorb ODS II, RP-C_{18}, 5 μm), using a slow water-methanol gradient as mobile phase; the detection occurred simultaneously at 230 and 290 nm, two wave lengths, in the range of which spirostaphylotrichin A (**1**) and the majority of the other spirostaphylotrichins isolated later, show typical absorption maxima.

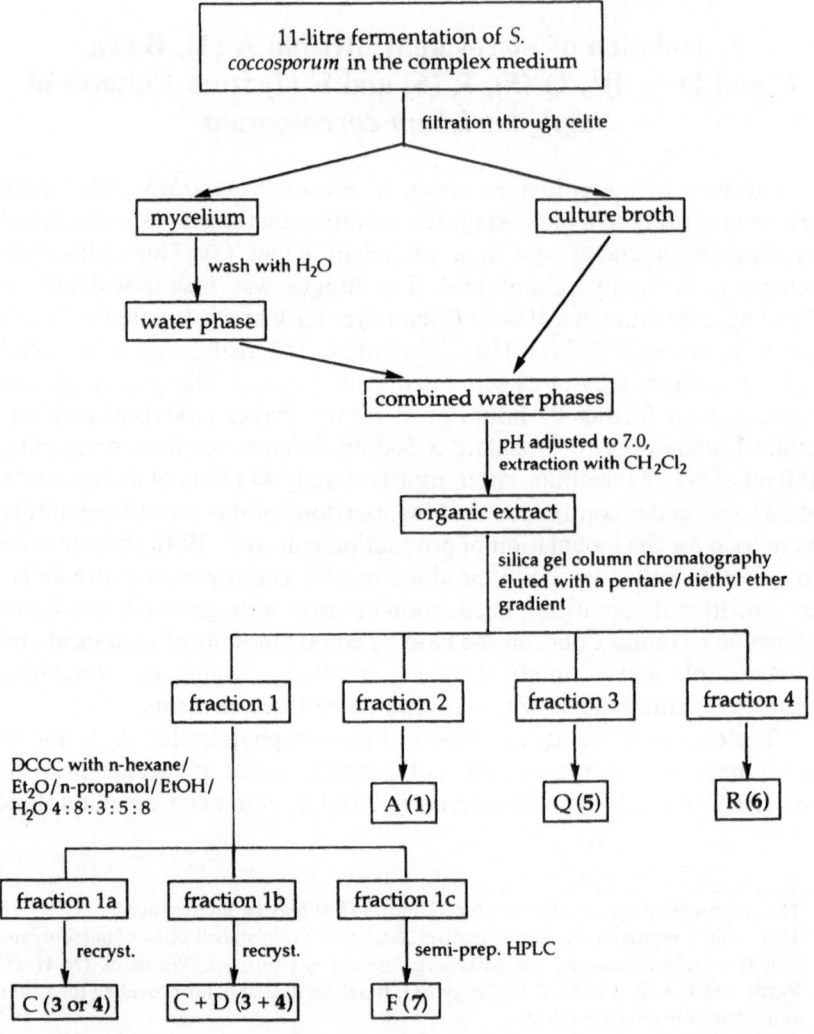

Fig. 1. Procedure of the isolation of metabolites starting from a culture of *S. coccosporum* in the complex medium

It was observed, that the cultures in the soya meal medium, as well as those in the minimal medium, reached maximum metabolite production after about 120–144 hours of fermentation, i.e. the mycelium stationary growth phase. Both kinds of production cultures showed at this fermentation stage, besides large amounts of spirostaphylotrichin A (1) (1–1.6 g per litre broths), present as the main metabolite, also smaller quantities of several unknown substances. From extracts obtained from a 11-litre fermentation in the complex medium, five compounds, namely spirostaphylotrichin A (1), C (3 or 4), Q (5), R (6) and F (7), were obtained in pure form, after being isolated following the procedure illustrated in Fig. 1. A sixth substance, spirostaphylotrichin D (4 or 3), was obtained as mixture with C, and all attempts to separate both compounds from each other were unsuccessful.

Starting from extracts of S. coccosporum cultures in the minimal medium, an additional new compound, called spirostaphylotrichin B (2), was subsequently isolated. Due to the close chemical similarity of B and A, the isolation of 2 required, after an initial open column chromatography of the crude extract on silica gel with pentane/ether, a supplementary purification step by semi-preparative HPLC (column: 16×250 mm, Lichrosorb Si 60, 7 µm; mobile phase: 1–2% methanol in CH_2Cl_2). Structure elucidations of all new spirostaphylotrichins mentioned in this section were based on the data of their IR, UV, EI-MS, ^{1}H-NMR and ^{13}C-NMR analysis, as well as on their comparison with those of A (1). A tabulation of the ^{1}H- and ^{13}C-NMR data is given in the Tables 1–4. Table 5 shows indeed some of the most important physical properties of A, B, C, Q and R.

Spirostaphylotrichin C and D were identified as the epimeric pair 3 and 4. A definite structural assignment was not possible on the basis of the measured spectroscopic data. Furthermore the attempt to characterize D was restricted to the ^{1}H-NMR examination of a mixture of both epimers. C and D structurally differ from A (1) in the presence of an oxo group at C(4) in place of the OH function in 1.

Spirostaphylotrichin B (2), the spectroscopic data of which show good correlation with those of A, turned out to be the 6-epimer of 1. Further investigation on B confirmed this structure 2 later. In the case of the spirostaphylotrichins Q (5) and R (6), SANDMEIER and TAMM demonstrated, that both compounds represent just artefacts, which are formed from A (1) during the work up procedure of the culture extracts. Because of the low stability of these substances and the very small amounts available, it was not possible to determine their configuration. However, the observation that A is partially converted to Q in a CDCl$_3$ solution containing traces of water, seems to suggest that this artefact has the

Table 1. ^1H-NMR Data ($CDCl_3$) of the Spirostaphylotrichins A(1), B(2), C and D (3/4), Q(5), and R(6)[a]

H-Atom	A(1)	B(2)	C(3 or 4)	D(3 or 4)[b]	Q(5)	R(6)
OH–C(3)						5.54 (s)
H–C(4)	4.64 (dt, $J = 6.2, 1.5$; with D_2Ot, $J = 1.5$)	5.06 (dt, $J = 66, 1.8$; with D_2Ot, $J = 1.8$)			4.02 (s)	4.03 (d, $J = 7$)
OH–C(4)	2.63 (d, $J = 6.2$)[c]	1.83 (d, $J = 6.5$)[c]				4.43 (d, $J = 7$)[c]
H–C(6)	4.76 (d, $J = 2.1$)[d]	4.25 (d, $J = 2.4$)[d]	4.74 (d, $J = 2.1$)[d]	4.65 (br. s)[d]	3.99 (br. s)[d]	4.74 (d, $J = 2.1$)[d]
OH–C(6)	3.80 (d, $J = 2.1$)[c]	3.97 (d, $J = 2.8$)[c]	3.71 (d, $J = 2.2$)[c]	4.0 (br. s)[c]	4.40 (br. s)[c]	2.2 (br. s)[c]
H–C(8)	5.93 (d, $J = 9.8$)	5.99 (d, $J = 9.8$)	6.07 (d, $J = 10.1$)	6.07 (d, $J = 10.1$)	5.79 (d, $J = 9.8$)	5.91 (d, $J = 9.8$)
H–C(9)	7.07 (d, $J = 10.1$)	7.07 (d, $J = 10.1$)	7.05 (d, $J = 10.1$)	7.05 (d, $J = 10.0$)	7.06 (d, $J = 10.4$)	7.04 (d, $J = 10.0$)
H–C(11)	4.72 (t, $J = 1.8$)	4.78 (t, $J = 1.8$)	5.27 (d, $J = 1.9$)	5.41 (d, $J = 2.1$)	1.64 (s, 3H)	1.63 (s, 3H)
H–C(11)	4.56 (t, $J = 1.7$)	4.63 (t, $J = 1.7$)	4.91 (d, $J = 1.9$)	4.95 (d, $J = 2.1$)		
H–C(12)	6.17 (t, $J = 7.4$)	6.20 (t, $J = 7.9$)	6.05 (t, $J = 7.0$)	6.05 (t, $J = 7.0$)	6.35 (t, $J = 7.9$)	6.14 (t, $J = 7.5$)
H_2–C(13)	2.1 (m)	2.2 (m)	2.1 (m)	2.1 (m)	2.36 (m)	2.29 (m)
H_3–C(14)	1.05 (t, $J = 7.3$)	1.02 (t, $J = 7.3$)	1.03 (t, $J = 7.4$)	0.99 (t, $J = 7.3$)	1.05 (t, $J = 7.3$)	1.04 (t, $J = 7.3$)
H_3–C(15)	3.92 (s)	3.88 (s)	4.06 (s)	3.96 (s)	3.86 (s)	3.99 (s)

[a] For all compounds, the numbering according to 1 is used.
[b] Data from the mixture 3/4.
[c] Exchangeable with D_2O.
[d] With D_2Os.

Table 2. ^{13}C-NMR Data of the Spirostaphylotrichins A (1), B (2), and Q (5)[a]

C-Atom	A (1) (CDCl$_3$)[b]	B (2) (CD$_3$OD/(D$_6$)DMSO)	Q (5) (CDCl$_3$)
C(1)	167.5 (169.8) (S)	167.8 (S)	171.2 (S)
C(3)	143.9 (145.5) (S)	145.4 (S)	97.8 (S)
C(4)	64.7 (65.5) (D)	69.7 (D)	77.6 (D)[c]
C(5)	57.3 (58.7) (S)	58.3 (S)	59.9 (S)
C(6)	73.8 (74.9) (D)	74.9 (D)	76.2 (D)[c]
C(7)	197.9 (198.5) (S)	197.4 (S)	192.3 (S)
C(8)	120.6 (122.1) (D)	121.4 (D)	121.2 (D)
C(9)	152.5 (152.9) (D)	153.5 (D)	152.1 (D)
C(10)	128.3 (130.2) (S)	129.3 (S)	124.3 (S)
C(11)	86.3 (86.5) (T)	86.1 (T)	12.6 (Q)[d]
C(12)	150.9 (150.2) (D)	151.8 (D)	150.8 (D)
C(13)	23.1 (24.0) (T)	25.6 (T)	24.7 (T)
C(14)	13.3 (13.6) (Q)	13.4 (Q)	13.4 (Q)[d]
C(15))	62.2 (62.6) (Q)	62.8 (Q)	64.5 (Q)

[a] For all compounds, the numbering according to 1 is used.
[b] Values in parenthesis in CD$_3$OD/(D$_6$)DMSO.
[c, d] May be interchanged.

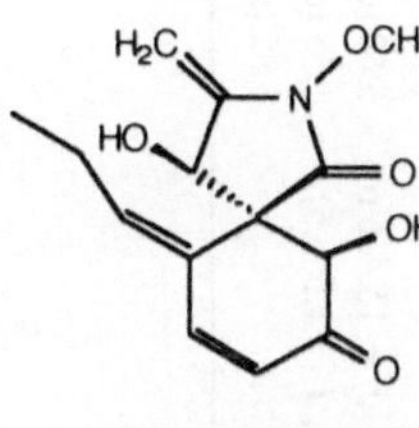

B (2)

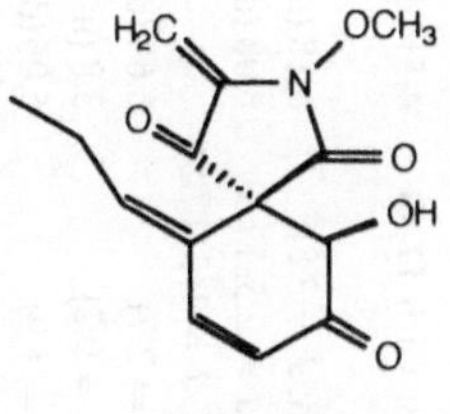

C/D (3)

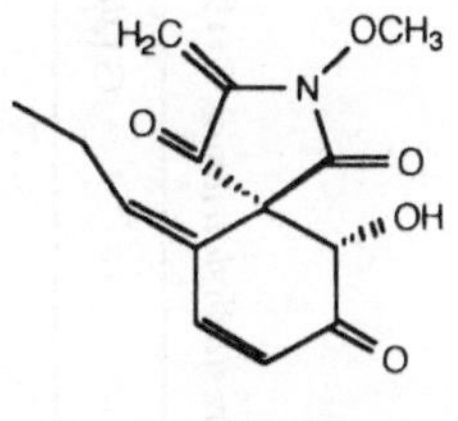

C/D (4)

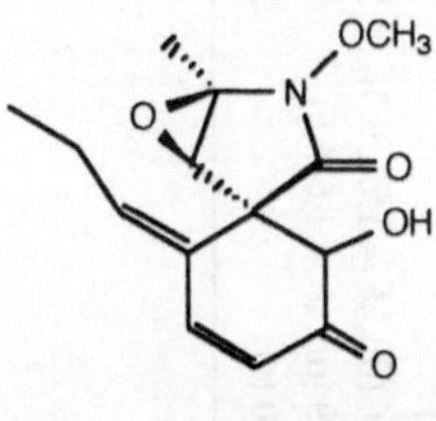

Q (5)

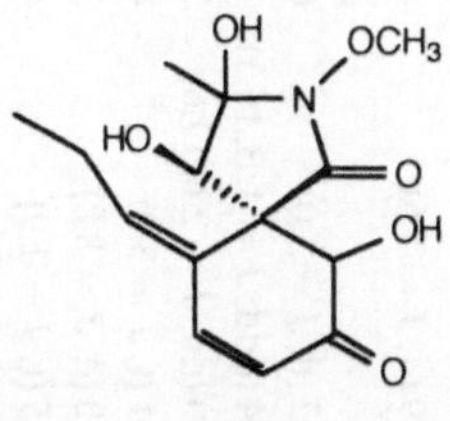

R (6)

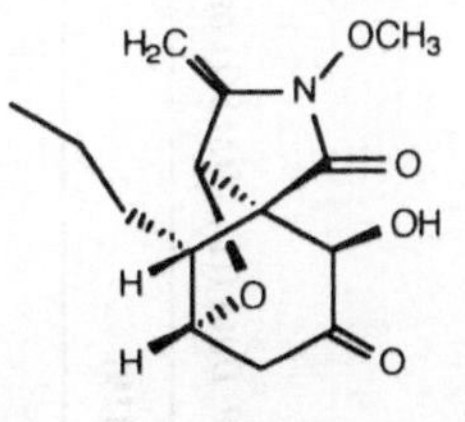

F (7)

same configuration as 1. Finally F (7) was isolated only in small amounts from the culture extracts of the wild type. Larger quantities of this compound, sufficient for the structure elucidation, were obtained from a mutant strain (P 84) of S. coccosporum (s. a. Chapter 3.2.2.).

Table 3. ^{1}H-NMR Data ((D_6)DMSO) of the Spirostaphylotrichins E (19), F (7), G and H (20/21), I (22), K (23), L (24), M (25), and S (26)ᵃ

H-Atom	E (19)	F (7)	G (20 or 21)	H (20 or 21)ᵇ	I (22)ᵇ
OH–C(3)					5.83 (s)ᵈ
H–C(4)	4.66 (dt, J = 1.8)	4.53 (t, J = 1.8)			4.09 (d, J = 6.0)
OH–C(4)					4.43 (d, J = 5.9)ᵈ
H–C(6)	4.49 (d, J = 5.5)ᶜ	4.18 (br. s; with D₂O sharper)	4.51 (d, J = 3.8)ᶜ	4.55 (d, J = 2.9)ᶜ	4.56 (d, J = 2.1)ᶜ
OH–C(6)	6.02 (d, J = 5.0)ᵈ	6.55 (br. d, J = 3.8)ᵈ	6.18 (d, J = 4.1)ᵈ	6.17 (d, J = 3.7)ᵈ	5.15 (d, J = 2.1)ᵈ
H–C(8)	2.74 (ddd, J = 1, 3, 16.3)	2.80 (dd, J = 2.5, 17.5)	6.08 (dd, J = 3.1, 10.1)	6.06 (dd, J = 2.8, 10.0)	6.26 (dd, J = 0.5, 9.8)
H–C(8)	2.50 (dd, J = 2.6, 17.6)	2.39 (ddd, J = 1, 3, 17.5)			
H–C(9)	4.41 (t, J = 3.0)	4.39 (t, J = 2.7)	6.68 (dd, J = 2.2, 10.1)	6.64 (dd, J = 2.2, 10.2)	6.93 (dd, J = 6.0, 9.8)
H–C(10)	3.30 (d, J = 8.8)	3.59 (d, J = 9.2)	3.8 (m)	3.8 (m)	3.64 (br. t, J = 5.9)
H–C(11)	4.55 (t, J = 1.9)	4.50 (t, J = 1.9)	5.06 (d, J = 2.2)	5.17 (d, J = 2.2)	1.57 (d, J = 1.2, 3H)ᶜ
H–C(11)	4.40 (t, J = 2.0)	4.34 (t, J = 1.7)	4.85 (d, J = 2.1)	4.88 (d, J = 2.3)	
H–C(12)	5.27 (ddq, J = 9.0, 15.3, 1.7)	5.28 (ddq, J = 9, 2, 15.4, 1.7)	5.13 (ddq, J = 8.8, 15.0, 1.5)	5.24 (ddq, J = 8.8, 15.3, 1.5)	5.8 (m)
H–C(13)	5.66 (dq, J = 15.5, 6.5)	5.66 (dq, J = 15.4, 6.6)	5.69 (dq, J = 15.1, 6.2)	5.58 (dq, J = 15.2, 6.5)	5.8 (m)
H₃–C(14)	1.59 (dd, J = 1.6, 6.5)	1.58 (dd, J = 1.5, 6.5)	1.58 (dd, J = 1.0, 6.2)	1.58 (d, J = 6.4)	1.76 (dd, J = 0.6, 5.0)
H₃–C(15)	3.70 (s)	3.69 (s)	3.85 (s)	3.81 (s)	3.87 (s)

Table 3 (*continued*)

H-Atom	K (23)	L (24)	M (25)	S (26)
OH–C(3)				5.29 (s)[d]
H–C(4)	4.96 (dt, J = 5,1; with D$_2$Ot, J = 1)	5.31 (br. s)	5.13 (t, J = 1.5)	4.28 (d, J = 6.2)[c]
OH–C(4)	6.35 (d, J = 5.1)[d]			5.84 (d, J = 6.4)[d]
H–C(6)	4.18 (d, J = 3.7)[c]	3.93 (br. d)[c]	3.87 (s)	4.12 (d, J = 3.7)[c]
OH–C(6)	5.61 (d, J = 3.7)[d]	5.10 (d, J = 4.3)[d]		5.42 (d, J = 3.7)[d]
H–C(7)		3.64 (m, with D$_2$O br. d, J = 5.4)		
OH–C(7)		4.99 (d, J = 2.5)	6.45 (s)	
H–C(8)	6.01 (dd, J = 3.2, 10.2)	1.83 (ddd, J = 1.8, 5.9, 15.0)	2.37 (dd, J = 2.8, 14.8)	5.98 (dd, J = 3.0, 10.3)
H–C(8)		1.76 (dd, J = 3.2, 15.0)	1.83 (dd, J = 1.6, 14.4)	
H–C(9)	6.70 (dd, J = 2.2, 10.2)	4.18 (br. s)	4.20 (br. t, J = 2.5)	6.68 (dd, J = 2.4, 10.3)
H–C(10)	3.82 (m)	3.17 (d, J = 9.2)	3.07 (d, J = 9.2)	3.7[e]
H–C(11)	4.41 (dd, J = 1,1)	4.43 (br. s)	4.43 (t, J = 1.5)	1.39 (s, 3H)
H–C(11)	4.34 (dd, J = 1,1)	4.30 (br. s)	4.31 (t, J = 1.5)	
H–C(12)	5.40 (ddq, J = 7, 16, 1)	5.27 (ddq, J = 9.2, 15.6, 1.5)	5.24 (ddq, J = 9.2, 15.2, 1.6)	5.73 (ddq, J = 7.2, 15.7, 1.6)
H–C(13)	5.51 (dq, J = 16, 5)	5.52 (dq, J = 15.4, 6.5)	5.54 (dq, J = 15.0, 6.6)	5.44 (dq, J = 15.7, 6.4)
H$_3$–C(14)	1.55 (dd, J = 1, 47)	1.54 (br. d, J = 6.2)	1.54 (dd, J = 1.3, 6.5)	1.61 (d, J = 7.0)
H$_3$–C(15)	3.68 (s)	3.68 (s)	3.68 (s)	3.70 (s)

[a] For all compounds, the numbering according to **1** is used.
[b] In CDCl$_3$/(D$_6$)DMSO 20:1.
[c] With D$_2$O, s.
[d] Exchangeable with D$_2$O.
[e] Submerged by the CH$_3$(15) signal.

Table 4. ^{13}C-NMR Data $((D_6)DMSO)$ of the Spirostaphylotrichins E (19), F (7), K (23), L (24), M (25), and S (26)[a]

C-Atom	E (19)	F (7)	K (23)	L (24)	M (25)	S (26)
C(1)	163.5(S)	164.0(S)	165.9(S)	166.3(S)	163.9(S)	165.0(S)
C(3)	141.3(S)	141.2(S)	144.7(S)	142.9(S)	142.1(S)	85.5(S)
C(4)	74.1(D)[b]	74.9(D)[b]	65.5(D)[b]	73.1(D)[b]	74.6(D)	70.5(D)[b]
C(5)	59.4(S)	57.0(S)	59.1(S)	56.9(S)	55.1(S)	56.9(S)
C(6)	75.1(D)[b]	75.3(D)[b]	72.2(D)[b]	73.7(D)[b]	70.0(D)	73.0(D)[b]
C(7)	207.7(S)	205.3(S)	197.0(S)	69.7(D)[b]	95.3(S)	197.3(S)
C(8)	47.4(T)	45.5(T)	127.4(D)[c]	36.4(T)	43,7(T)	127.4(D)[c]
C(9)	82.1(D)	81.9(D)	148.1(D)	82.5(D)	81.2(D)	148.7(D)
C(10)	52.0(D)	46.8(D)	40.5(D)	47.1(D)	46.9(D)	39.9(S)
C(11)	84.5(T)	83.9(T)	81.7(T)	82.6(T)	82.5(T)	24.9(Q)
C(12)	125.6(D)[c]	126.2(D)[c]	126.4(D)[c]	127.5(D)[c]	126.7(D)	125.9(D)[c]
C(13)	129.5(D)[c]	129.3(D)[c]	129.1(D)[c]	128.2(D)[c]	128.8(D)	130.3(D)[c]
C(14)	17.6(Q)	17.7(Q)	17.7(Q)	17.6(Q)	17.6(Q)	17.8(Q)
C(15)	62.1(Q)	62.0(Q)	61.9(Q)	61.9(Q)	61.8(Q)	64.1(Q)

[a]   For all compounds, the numbering according to 1 is used.
[b,c]   May be interchanged.

Table 5. Physical Properties of the Spirostaphylotrichins B, C, Q, and R

Compound	A (1)	B (2)	C (3 or 4)
Appearance	Colourless crystals	Colourless crystals	Slightly yellow crystals
Melting point	106–111°	162–176°	133–138°
Molecular formula	$C_{14}H_{17}O_5N$	$C_{14}H_{17}O_5N$	$C_{14}H_{15}O_5N$
UV(EtOH)	224(19300),	223(19000),	222(13700),
λ_{max} nm(ε)	289(15400)	290(13600)	284(14300)
IR ν_{max}(KBr) cm^{-1}	3460 (br), 3390 (br), 1720s, 1680s, 1660s, 1615m, 1260m, 1140m, 1110m, 1070m	3410 (br), 3320 (sh), 1720s, 1670s, 1650s, 1620s, 1575m, 1260m, 1075m	3440s, 2980w, 1765m, 1735s, 1685s, 1645m, 1620m, 1285m, 1110m, 955m

Compound	Q (5)	R (6)
Appearance	Colourless crystals	Colourless crystals
Melting point	163–169°	157–166°
Molecular formula	$C_{14}H_{17}O_5N$	$C_{14}H_{17}O_5N$
UV(EtOH)	205(8900), 291(14200)	204(9100), 291(14100)
λ_{max} nm(ε)		
IR ν_{max}(KBr) cm^{-1}	3400s, 3000w, 2980w, 2940w, 1720s, 1680s, 1630m, 1585m, 1235m, 1115m, 980m	3480 (br), 3360 (br) 1700s, 1675s, 1615m, 1580m, 1410m, 1215, 1185m, 1120m, 1105m, 1075m, 975m

3. Biosynthetic Studies

3.1. Feeding Experiments with ¹⁴C-, ¹³C- and ²H-Labelled Precursors

3.1.1. Experiments with ¹⁴C-Labelled Precursors

In order to determine the building blocks involved in the biosynthesis of spirostaphylotrichin A (1), a series of incorporation experiments with potential labelled precursors was carried out. Since it seemed reasonable to assume for this γ-lactam metabolite a mixed biosynthetic origin resulting from the condensation of a polyketide chain with an amino acid and involving the participation of L-methionine (or more exactly of S-adenosyl-L-methionine) as probable donor of the 15-methyl group, the following ¹⁴C-labelled precursors were chosen for initial investigations: sodium [1-¹⁴C]- and [2-¹⁴C]acetate, L-[methyl-¹⁴C]methionine, DL-[1-¹⁴C]alanine, DL-[1-¹⁴C]serine and L-[U-¹⁴C]aspartic acid. All feeding experiments were conducted with cultures of *S. coccosporum* (*wild type*) in both kinds of production media (*3*). In general, precursors were administered to the fermentation in single doses and at a time on which

Table 6. *Incorporation Experiments with ¹⁴C-Labelled Precursors*

Precursor	Total precursor activity [μCi]	Specific precursor activity [mCi/mmol]	Time[a] [h]	Medium[b]	Absolute incorporation rate [%]
Sodium [1-¹⁴C]acetate	14.3	59.1	0	S	2.8
Sodium [1-¹⁴C]acetate	14.3	59.1	24	M	14.8
Sodium [2-¹⁴C]acetate	17.2	55.0	0	S	2.9
L-[Methyl-¹⁴C] methionine	13.1	51.0	0	S	7.6
L-[Methyl-¹⁴C] methionine	13.1	51.0	24	S	21.1
DL-[1-¹⁴C]Alanine	14.1	47.0	24	S	0.52
DL-[1-¹⁴C]Alanine	14.1	47.0	24	M	0.58
DL-[1-¹⁴C]Serine	6.1	51.0	24	S	0.36
DL-[1-¹⁴C]Serine	6.1	51.0	24	M	0.12
L-[U-¹⁴C]Aspartate	6.4	200	24	S	1.2
L-[U-¹⁴C]Aspartate	6.4	200	24	M	4.0
L-[U-¹⁴C]Aspartate[c]	6.4	$8.5 \cdot 10^3$	24–48	M	5.6

[a] Addition of the precursors to the cultures.
[b] S: soya meal medium; M: minimal medium.
[c] L-[U-¹⁴C]aspartate diluted with 100 mg of unlabelled material and added in 5 portions between 24 and 48 h.

production of spirostaphylotrichins was just detectable in the culture broth. In the particular case of L-[U-^{14}C]aspartic acid a pulse feeding protocol was also followed with the aim to increase the incorporation rate. The results of these feeding experiments with ^{14}C-labelled precursors are summarized in Table 6. They provided important information to guide successive studies with stable isotopic materials.

The extensive incorporation rates of acetate, particularly significant if one considers that this precursor is very unspecific, seemed to support the polyketide hypothesis mentioned above. High incorporation was also observed for L-[methyl-^{14}C]methionine. More complicated however, was the interpretation of the results of the feeding experiments carried out with the three ^{14}C-labelled amino acids. Indeed, the incorporation rates of DL-[1-^{14}C]alanine, DL-[1-^{14}C]serine and L-[U-^{14}C]aspartic acid are clearly superior to the significance limit of 0.1%, but are also relatively low as compared for instance with the rates observed in case of 19-O-acetylchaetoglobosin A for DL-tryptophan (11.7%) (*11*), or pseurotin A for L-phenylalanine (10%) (*12*). In any case, the most probable biosynthetic precursor among these three amino acids appeared to be L-aspartate. In the case of alanine and serine it was reasonable to assume that they were not incorporated as intact units, but *via* their degradation to acetate.

3.1.2. Experiments with ^{13}C- and ^{2}H-Labelled Precursors

Having established the basic biogenetic units of spirostaphylotrichin A (**1**), the various precursors were also fed as ^{13}C-labelled molecules in order to elucidate the labelling pattern. Administration of sodium [1-^{13}C]acetate to a culture of *S. coccosporum* in the complex production medium, according to a pulse feeding protocol, led to enhanced signals in the ^{13}C-NMR spectrum of the resulting spirostaphylotrichin A (**1**) (relative to the ^{13}C-NMR of unlabelled **1**, (Fig. 2a)) for C(1), C(4), C(6), C(8), C(10) and C(13) (Fig. 2b). The calculated enrichment factors (EF) amounted to 0.8% for C(4) and to 1.5–2.8% in the case of the other labelled positions. An analogous feeding experiment with [2-^{13}C]acetate gave rise to enhanced signals with average enrichments of *ca.* 2.4% for C(5), C(7), C(9), C(11), C(12) and C(14), and of about 0.8% for C(4) and C(15). The latter feeding experiment was repeated under similar conditions with a production culture of *S. coccosporum* in the minimal medium. In this case higher enrichments were observed for C(5), C(7), C(9) and C(14) (average EF of *ca.* 3.0%), only modest enrichments for C(3), C(11) and C(12) (EF *ca.* 1.1%) and no significant incorporation into

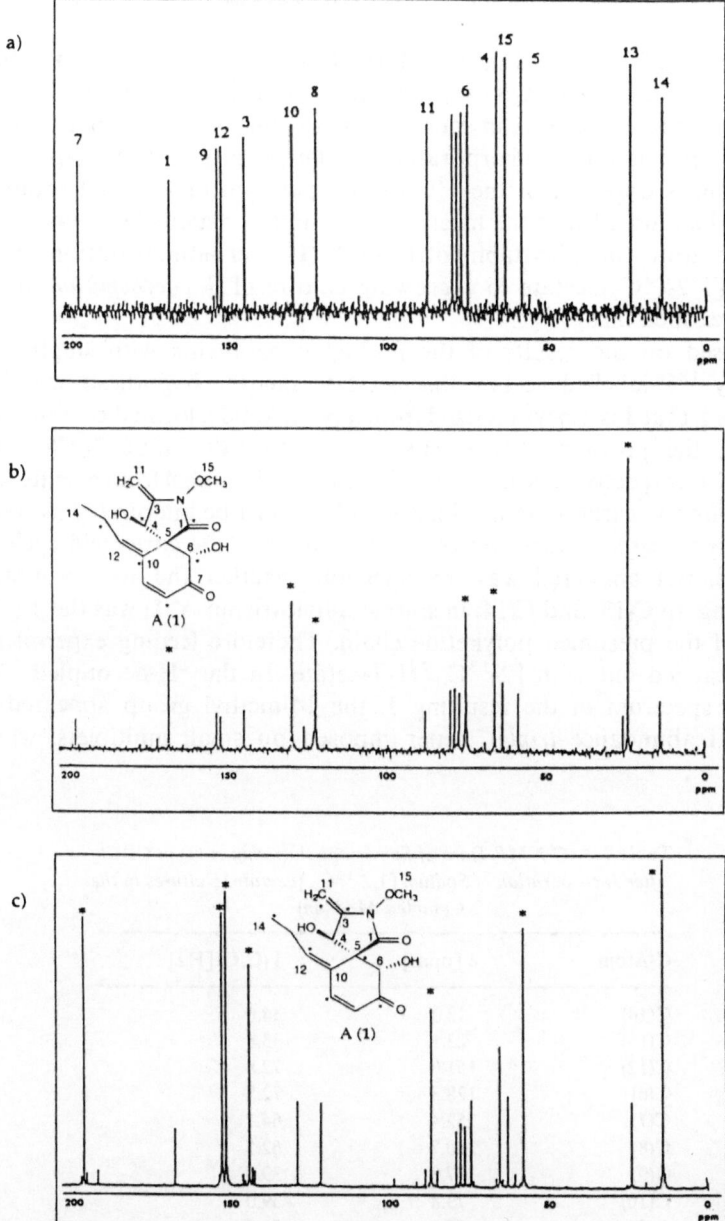

Fig. 2. Noise-decoupled 13C-NMR spectrum of spirostaphylotrichin A (**1**) in CDCl₃; a: under normal conditions, b: after incorporation of [1-¹³C]acetate (fermentation in the complex medium) and c: after incorporation of [2-¹³C]acetate (fermentation in the minimal medium)

C(4) and C(15) (Fig. 2c). Further information regarding the incorporation of intact acetate moieties was obtained from experiments with [1, 2-$^{13}C_2$]acetate. In general, in such an experiment, only those positions in which there is intact incorporation of the precursor, strong spin-spin coupling is observed in the ^{13}C-NMR spectrum of the final product. Table 7 shows a list of the measured coupling constants between linked carbon atoms in spirostaphylotrichin A (1) after administration of sodium [1, 2-$^{13}C_2$]acetate to a growing culture of S. coccosporum in the complex medium.

Based on the results of the feeding experiments with singly and doubly ^{13}C-labelled acetate discussed so far, the hypothesis was formulated, that 1 is biosynthesized from a pentaketide, located corresponding to the positions C(14)/C(13), C(12)/C(10), C(9)/C(8), C(7)/C(6) and C(5)/C(1) respectively (s.a. Scheme 1). The fact that label from acetate was also found in carbon atoms No. 3, 4 and 11 must be interpreted as result of a high degree of turnover of this precursor in the citric acid cycle.

Still not answered was the question, whether the acetate moiety building up C(13) and C(14) in spirostaphylotrichin A (1) was the starter unit of the presumed polyketide chain. Therefore feeding experiments were carried out with [2-^{13}C, $^2H_3$]acetate. In the ^{1}H-decoupled ^{13}C-NMR spectrum of the resulting 1, the 14-methyl group appeared as natural abundance *triplet*, super-imposed on small multiplets, which

Table 7. ^{13}C-NMR Data of Spirostaphylotrichin A (1) in CDCl$_3$ After Incorporation of Sodium [1,2-$^{13}C_2$]acetate (Cultures in the Complex Medium)

C-Atom	δ [ppm]	J (C, C) [Hz]
C(14)	13.3	34.6
C(13)	23.1	33.8
C(12)	151.0	72.8
C(6)	128.3	72.8
C(7)	152.4	64.7
C(8)	120.7	62.5
C(9)	197.0	39.0
C(10)	73.8	39.0
C(5)	57.3	50.0
C(1)	167.4	50.0
C(4)	64.8	49.3
C(3)	143.9	49.3
C(11)	86.2	83.8
C(15)	62.2	–

were identified as a *triplet* (CDH_2) with an isotopic shift of 0.28 ppm to high field, a *quintet* (CD_2H) with a shift of 0.55 ppm and part of a *septet* (CD_3) shifted by 0.83 ppm. These results plainly confirmed that C(13) and C(14) are the starter unit of the polyketide chain involved in the biosynthesis of **1**.

L-methionine, which after the feeding experiments with radioactive precursors had been supposed to be the C_1-donor of the 15-methyl group of **1**, was administered with the methyl group labelled by ^{13}C. In the ^{13}C-NMR spectrum of spirostaphylotrichin A (**1**), obtained after the administration of L-[methyl-^{13}C]methionine to the production cultures, the intensity of C(15) was increased about 26-fold, confirming the participation of this precursor in the spirostaphylotrichin biosynthesis.

Finally, the uncertain origin of the three C-atoms No. 3, 4 and 11 remained yet to be examined in more detail. As already mentioned earlier, preliminary pulse feeding experiments with L-[U-^{14}C]aspartate in production cultures in the minimal medium, showed rather good incorporation of this precursor in spirostaphylotrichin A (**1**). On the other hand, the label originating from different ^{13}C-labelled acetate precursors tested, unexpectedly was found to be distributed, apart into those positions belonging to the polyketide, also on C(3), C(4) and C(11). The combination of these results suggested the hypothesis that these three carbon atoms of the lactam ring are derived from L-aspartic acid or from a C_4-intermediate of the citric acid cycle. Therefore, further feeding experiments with L-[2,3-$^{13}C_2$]aspartic acid and [2,3-$^{13}C_2$]succinate were carried out. Both precursors were synthesized using appropriately labelled starting materials (*3*). Particularly interesting in this context was the synthesis of L-[2,3-$^{13}C_2$]aspartic acid, in which the last step concerning the conversion of [2,3-$^{13}C_2$]fumaric acid into the desired labelled amino acid, was performed by immobilized cells of *E. coli* (ATCC11303) (*13, 14*). The ^{13}C-NMR spectrum of enriched spirostaphylotrichin A (**1**), isolated after administration of L-[2,3-$^{13}C_2$]aspartic acid to growing cultures in the minimal medium, showed high incorporation into C(3) and C(11) ($J_{3, 11} = 83.8$ Hz, EF = 1.99%). This result appears to be consistent with the assumption that C(3), C(4) and C(11) are derived from this amino acid. In the above mentioned ^{13}C-NMR spectrum also coupling signals were observed of lower intensity for such carbon atoms which have previously been found to originate from acetate, thus demonstrating that the labelled aspartate had been extensively metabolized. The results of feeding experiments with [2,3-$^{13}C_2$]succinate turned out to be very similar to those just discussed, with the exception of lower enrichment factors for C(3) and C(11) (EF = 1.83%) and of a higher degree of scrambling of label in other positions. For this reason it was not possible to determine unambiguously the C_4-unit leading to $C(3)$, $C(4)$ and $C(11)$ in **1**. However, it appears reasonable to consider L-aspartic acid as being the most probable direct biosynthetic precursor, also not at least because it would also explain the origin of the nitrogen atom of the lactam function. A possible proof of the correctness of this assumption could be provided by intact incorporation of L-$[2-^{13}C,^{15}N]$aspartate into **1**. But this feeding experiment was not carried out because of its poor chance of success. Indeed, it is to be expected, that the major part of the N-label would be lost due to the important role that aspartate plays in transamination reactions.

Summarizing the results of the incorporation experiments discussed, the C-skeleton of spirostaphylotrichin A (**1**) has been shown to arise from the condensation of a pentaketide with a C_4-dicarboxylic acid, most probably L-aspartic acid, which is biosynthetically linked to the citric acid cycle (Scheme 1). The 15-methyl group originates from L-methionine (s.a. Chapter 3.2.3.). The pentaketide is biosynthesized starting from acetyl-coenzyme A, localized in spirostaphylotrichin A (**1**) at $C(14)/C(13)$. This starting unit is condensed with four malonyl-coenzyme A units, in a series of chain elongation steps being part of the complicated reaction programme catalysed and coordinated by the polyketide synthase multienzyme complex.

The reaction sequence in Scheme 1 illustrates the conclusion mentioned above in a quite simplified fashion. The poly-β-ketone **8** may rather be considered as pure formalism, instead of the real intermediate of the spirostaphylotrichin biosynthesis. It is indeed improbable, that the biosynthesis of the C_{10}-polyketide precursor of **1** exclusively comprises of a succession of four chain extending reactions with malonyl-coenzyme A and no reduction of keto groups, dehydration and enone reduction step as it is the case in fatty acid biosynthesis (*15, 16*). Recent studies on the biosynthesis of polyketide metabolites, conducted on different com-

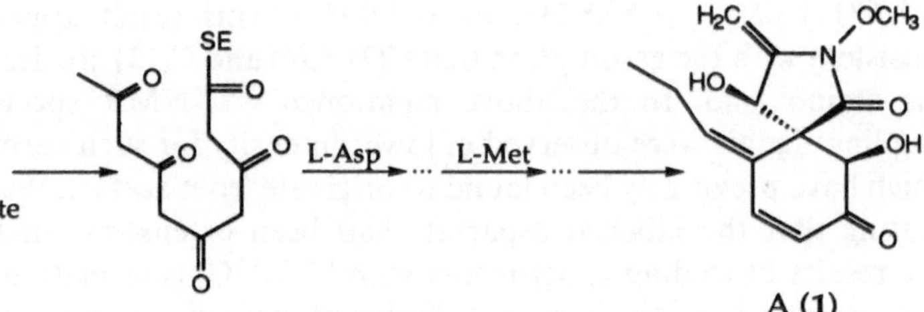

Scheme 1

pounds such as the antibiotics erythromycin A (or rather on its aglycone 6-deoxyerythronolide B) (17–19), methymycin (20), tylactone (21), narginin (22), monensin A (23) and tetronasin (24–26), have established that polyketide formation often proceeds according to a "processive" mechanism. In these cases, the oxidation level and stereochemistry of the growing polyketide chain are adjusted subsequently to each condensation step and prior to the addition of the next elongation unit. If this biosynthetic principle is also valid for 1, then we could suggest for the

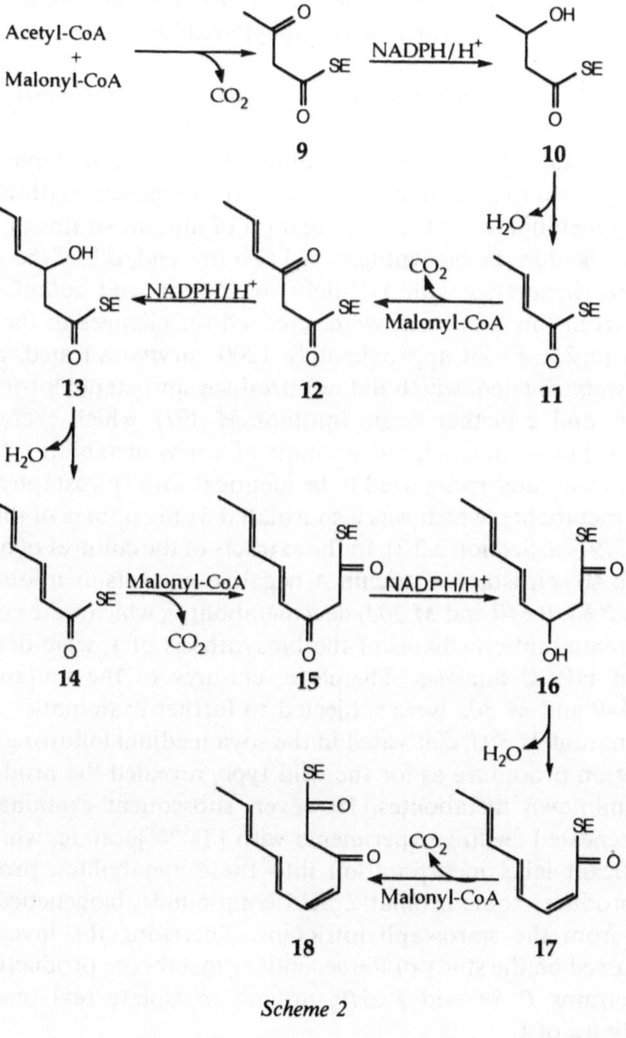

Scheme 2

biosynthesis of the spirostaphylotrichin pentaketide the reaction sequence illustrated in Scheme 2. Evidence confirming this hypothetical pathway could eventually be obtained by the specific incorporation of several of the polyketide chain elongation intermediates shown in Scheme 2, appropriately labelled and activated as acetylcysteamine thioesters, into spirostaphylotrichin A (1) following the method developed by CANE (*17, 18, 20, 22*) and HUTCHINSON (*21*). This incorporation method however has been applied so far exclusively to streptomycetes.

3.2. Investigation of Mutant Strains and Isolation of Further Spirostaphylotrichins

3.2.1. Mutagen Treatment and Selection of Mutants

Having established the basic building blocks of spirostaphylotrichin A (1), it was tried to gain more insight into the biogenetic pathway of this secondary metabolite *via* the investigation of mutant strains of *S. coccosporum* blocked in the biosynthesis of 1. To this end, cell of the wild type strain were irradiated with UV-light at 254 nm and colonies having survived from this treatment were screened for changes in the product spectrum (*5, 27*). From approximately 1200 survivors tested, about 40 mutants were isolated, which did not produce spirostaphylotrichin A (1) any more, and a further strain (mutant *M 391*), which excreted only traces of 1, but relatively large amounts of a new metabolite. The latter compound was later recognized to be identical with spirostaphylotrichin O (29), a metabolite, which was also isolated from cultures of the mutant strain *P 649* (s.a. Section 3.2.3). In the extracts of the cultures of only three out of the 40 spirostaphylotrichin A negative mutants mentioned above (mutants *P 84*, *P 649* and *M 303*) new metabolites, which were considered to be potential intermediates of the biosynthesis of 1, were detected by TLC and HPLC analysis. Therefore, cultures of the mutant strains *P 84*, *P 649* and *M 303* were subjected to further examination.

The mutant *M 303*, cultivated in the soya medium following a similar fermentation procedure as for the wild type, revealed the production of several unknown metabolites. However, subsequent examination, including repeated feeding experiments with [1-^{14}C]acetate, which led to no significant label incorporation into these metabolites, proved that *M 303* produces some aromatic, acid compounds, biogenetically independent from the spirostaphylotrichins. Therefore, the investigations were focussed on the study of the secondary metabolite production of the mutant strains *P 84* and *P 649*, hoping to isolate real biosynthetic intermediates of 1.

3.2.2. The Mutant Strain P 84

The mutant strain *P 84* was cultivated in 11-litre fermentations using the complex, as well as the minimal medium. HPLC analysis of the respective culture broths revealed a different pattern of products: several compounds appeared to be present in both kind of cultures (*e.g.* spirostaphylotrichin E (**19**) and F (**7**)), and others only in one of the two production media. The details on the distribution of these products in the two different fermentations of *P 84* are reported in a previous paper (*5*). Culture broths were extracted separately with diethylether or CH_2Cl_2 and the concentrated organic solutions were subjected to repeated middle-pressure chromatography on silica gel with different solvents, yielding nine new substances, namely the spirostaphylotrichins E (**19**), F (**7**), G (**20**, or **21**), H (**21** or **20**), I (**22**), K (**23**), L (**24**), M (**25**) and S (**26**).

E (**19**)   G/H (**20**)   G/H (**21**)

I (**22**)   K (**23**)   L (**24**)

M (**25**)   S (**26**)

The structures of these nine spirostaphylotrichins were elucidated by spectroscopic methods, in some cases complemented by chemical modifications and conversions (5). The ^{1}H- and ^{13}C-NMR data of compounds 7 and 19–26 are summarized in Tables 3 and 4. Their physical properties are shown in Table 8.

All compounds isolated from the cultures of *P 84* appear to be structurally closely related to spirostaphylotrichin A (1). They have in common the (*E*)-prop-1-enyl side chain, in place of the propylidene side chain with Z-configuration, the latter being characteristic for all substances found in culture broths of the wild type strain, except for F (7) (cf. Chapter 2).

The relative configuration of spirostaphylotrichin K (23) was established by X-ray diffraction and ^{1}H-NMR-NOE difference experiments. Considering these results, as well as the fact that this compound is converted by selective hydrogenation to the same tetrahydrospirostaphylotrichin 27, which is also obtained starting spirostaphylotrichin B (2) (Scheme 3), the configuration as shown in formula 23 is proposed for spirostaphylotrichin K.

In the case of spirostaphylotrichin F (7), the assignment of configuration was performed on the basis of the relative configuration ascertained by X-ray diffraction and NOE experiments, and on the hypothesis (confirmed several times), that all spirostaphylotrichins possess the same configuration at the spiro centre as observed for 1. Spirostaphylotrichin E, whose spectroscopic data are in good agreement with those of F (7), turned out to be the 6-epimer of 7. This conclusion was confirmed by the strong ^{1}H-NMR-NOE observed for E between H-C(6) and H-C(10).

The spirostaphylotrichins G and H (20 and 21 or 21 and 20)[4] differ from the other metabolites isolated from cultures of the strain *P 84* by the presence of an oxo group at C(4), instead of the OH group, as for K (23), or of the O-bridge from C(4) to C(9), as in the case of F (7) and E (19). Although it was not possible so far to determine the absolute configuration of G and H, it seems reasonable to assume, that these compounds, which show large spectroscopic similarities, are also epimeric at C(6). For M the stereochemistry shown in structure 25 is suggested on the basis of the results of NOE difference experiments and of calculations using the MM2 programme (28), with the aim to determine the configuration with the most favorable energetic value. However, no valid proposals were possible with regard to the absolute configuration of the remaining spirostaphylotrichins I (22), L (24) and S (26).

---

[4] A final structural assignment was not possible on the basis of the spectroscopic data of both compounds. The structures can be interchanged.

Table 8. *Physical Properties of the Spirostaphylotrichins F, E, G, H, I, L, M and S*

Compound	F (7)	E (19)	G (20 or 21)
Appearance	Colourless crystals	Colourless crystals	Colourless crystals
Melting point	114–116°	162–176°	160–176°
Molecular formula	$C_{14}H_{17}O_5N$	$C_{14}H_{17}O_5N$	$C_{14}H_{15}O_5N$
UV(EtOH) $\lambda_{max}\,nm(\varepsilon)$	225 (12200)	225 (12200)	202 (10800), 232 (10000), 279 (3200)
IR ν_{max} (KBr) cm^{-1}	3420 (br), 3020w, 2870m, 2940w, 1730s, 1720s, 1695m, 1660s, 1445m, 1270m, 1210m, 1115m, 1035m, 975m, 850m, 840m	3400 (br., sh), 3350 (br), 3020w, 2980m, 2950m, 2920m, 2860w, 1745s, 1735s, 1670m, 1440m, 1380m, 1280m, 1145m, 1025m, 970m	3450 (sh), 3410 (br), 3010w, 2970w, 2860w, 1760m, 1725s, 1700s, 1650m, 1445m, 1315m, 1280m, 1120m, 970m, 960m, 905m, 865m

Compound	H (21 or 20)	I (22)	K (23)
Appearance	Colourless crystals	Colourless crystals	Colourless crystals
Melting point	–	167–176°	200–205°
Molecular formula	$C_{14}H_{15}O_5N$	$C_{14}H_{19}O_6N$	$C_{14}H_{17}O_5N$
UV(EtOH) $\lambda_{max}\,nm(\varepsilon)$	< 200, ca. 220 (sh, 7600), ca. 240 (sh, 5500)	< 200, ca. 220 (sh, 7600), 240 (sh, 5500)	226 (15100)
IR ν_{max} (KBr) cm^{-1}	3470 (br), 2940m, 2880m, 1770m, ca. 1730s (several), 1650m, 1310m, 1280m, 1120m, 980m	3480 (br), 3340 (br), 2940w, 2920w, 1695s, 1470m, 1240m, 1165m, 1095m, 960m	3420 (br), 3260 (br), 2945w, 2920w, 1715s, 1700s, 1675s, 1440m, 1275m, 1160m, 1110m, 1100m, 1080m, 995m, 975m

Compound	L (24)	M (25)	S (26)
Appearance	Colourless crystals	Colourless crystals	Colourless crystals
Melting point	131–137°	187–190°	166–173°
Molecular formula	$C_{14}H_{19}O_5N$	$C_{14}H_{17}O_5N$	$C_{14}H_{19}O_6N$
UV(EtOH) $\lambda_{max}\,nm(\varepsilon)$	228 (9500)	230 (5200)	< 200, 210–220 (sh, 7000), 230–240 (sh, 15100)
IR ν_{max} (KBr) cm^{-1}	3440 (br), 2950m, 1725s, 1695m, 1665s, 1270m, 1060m, 1025m, 975m	3400 (br), 2950m, 1730s, 1690s, 1670s, 1275m, 1120m, 1070s, 980m, 970m	3350 (br), 2950w, 1700s, 1410m, 1225m, 1100m

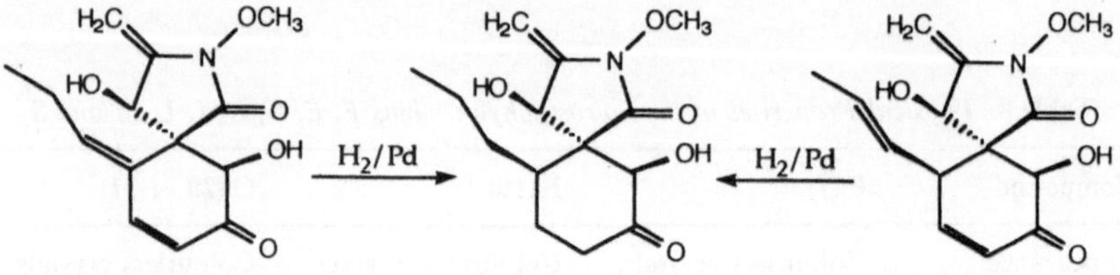

B (2)   27   K (23)

Scheme 3

3.2.3. The Mutant Strain P 649

In the course of some preliminary investigations it was observed, that the mutant strain *P 649* exhibits satisfactory metabolite production only in cultures in the minimal medium (*6*). For this reason *P 649* was cultivated, in view of subsequent investigation, in a fermenter containing 11 litre of the production medium just mentioned and following the same procedure as described for the wild type in Chapter 2. The fermentation was regularly controlled by HPLC and interrupted after about 105 hours. The culture broth was filtered and extracted several times with CH_2Cl_2. The crude extract, obtained after evaporating the solvent, was

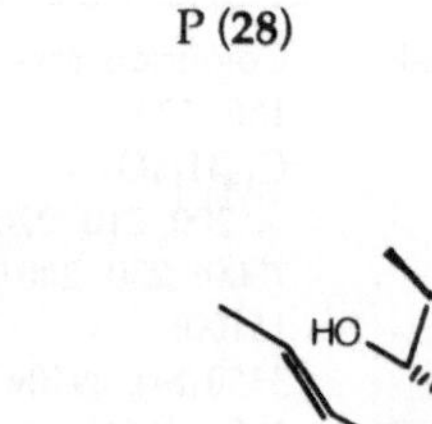

P (28) 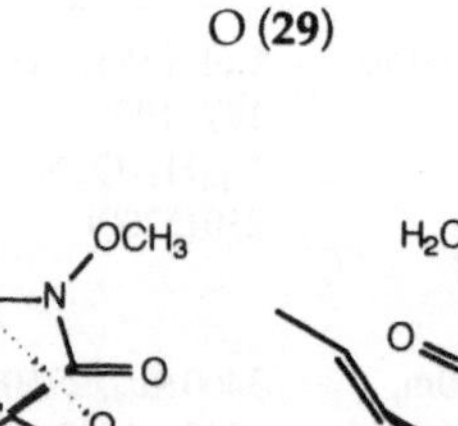   O (29) 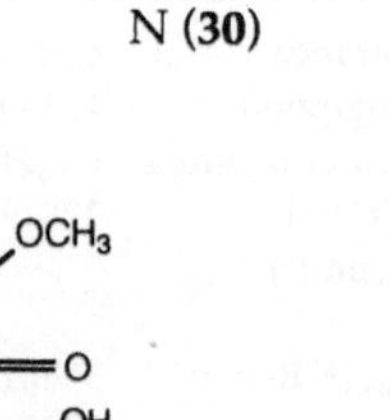   N (30)

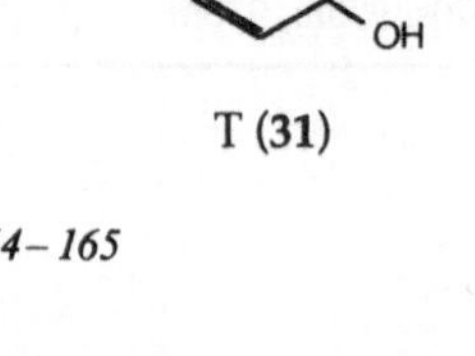

T (31) 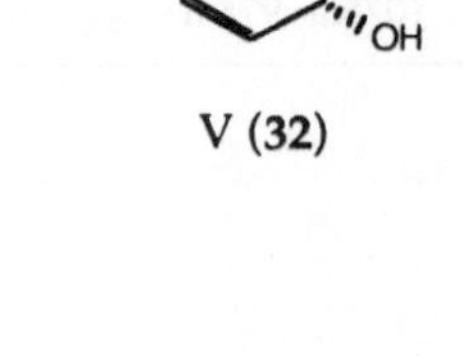   V (32)

digested with pentane to remove the lipids and afterwards submitted to repeated middle-pressure chromatography on silica gel with different mobile phases. In this way four new compounds were isolated, namely the spirostaphylotrichins P (**28**), O (**29**), N (**30**) and T (**31**). A fifth compound, spirostaphylotrichin V (**32**), was obtained in pure form by semi-preparative HPLC of a fraction containing a mixture of **32** and N (**30**) (*27*).

The structure elucidation of these five new compounds, whose physical properties are summarized in Table 9, was carried out on the basis of the data of NMR- and mass spectroscopy and by their comparison with those of already characterized spirostaphylotrichins (Table 10 and 11). A characteristic structural feature common to all spirostaphylotrichins

Table 9. *Physical Properties of the Spirostaphylotrichins P, O, N, T and V*

Compound	P (**28**)	O (**29**)	N (**30**)
Appearance	Colourless crystals	Colourless crystals	Colourless crystals
Melting point	141–144°	147–151°	92–95°
Molecular formula	$C_{13}H_{15}O_4N$	$C_{14}H_{19}O_5N$	$C_{14}H_{17}O_5N$
UV(EtOH) λ_{max} nm(ε)	222(8800), 275 (4000)	202(11200), 228 (12200)	225(7800), 275(4300)
IR ν_{max} (KBr) cm^{-1}	3500–3250 (br, several), 3040w, 2995w, 2975w, 2965w, 1760m, 1700s, 1660s, 1365m, 1300m, 1085m, 970m	3420 (br), 3040w, 2950w, 1665s, 1440m, 1270m, 1170m, 1090m, 1060m, 985m, 920m, 870m, 850m	3520m, 3400 (br), 3050w, 3005w, 2985w, 2975w, 2905w, 1750s, 1725s, 1650s, 1445m, 1320m, 1290m, 1110m, 970m

Compound	T (**31**)	V (**32**)
Appearance	Colourless crystals	Colourless crystals
Melting point	152–159°	–
Molecular formula	$C_{14}H_{19}O_5N$	$C_{14}H_{17}O_5N$
UV(EtOH) λ_{max} nm(ε)	202(7400)	228, 275
IR ν_{max} (KBr) cm^{-1}	3480 (br), 3270(br), 3040m, 2950m, 1730s, 1405m, 1190m, 1095m, 1065m, 995m, 880m	3540–3120(br), 3080m, 3040w, 2990w, 2890w, 1760s, 1730s, 1700s, 1520m, 1420m, 1340m, 1230m, 1110m, 975m

Table 10. ^1H-NMR Data ((D_6)DMSO) of the Spirostaphylotrichins P (**28**), O (**29**), N (**30**), T (**31**) and V (**32**)[a]

H-Atom	P (**28**)[b]	O (**29**)	N (**30**)	T (**31**)	V (**32**)
H-C(4)		5.07(m, with D_2O br. s)		3.84(br. s)[e]	
OH-C(4)		4.92(d, J = 6.2)[c]		5.92(br. d, J = 3.0)[c]	
H-C(6)	3.92(t[d], J = 5.0 with D_2O d, J = 4.8)	3.65(t, J = 5.5, with D_2O d, J = 5.1)	3.93(dd, J = 5.6)	3.69(d, J = 8.8)	3.95(dd, J = 5.0, with D_2O d, J = 5.5)
OH-C(6)	5.21(d, J = 5.1)[c]	6.25(d, J = 5.5)[e]	5.37(d, J = 5.8)[c]		5.43(d, J = 5.2)[c]
H-C(7)	3.88(m, with D_2O br. t, J = 4.8)	3.91(dt, J = 11.7, 4.8, with D_2O t, J = 4.5)	4.02(m)	3.53(m, with D_2O dt, J = 8.8, 1.5)	3.94(m, with D_2O dd, J = 4.0, 4.0)
OH-C(7)	4.12(d, J = 11.1)[c]	4.64(d, J = 11.6)[c]	3.63(d, J = 8.8)[c]	5.28(d, J = 5.9)[c]	3.75(d, J = 10.9)[c]
H-C(8)	5.89(ddd, J = 3.0, 4.8, 10.1)	5.84(ddd, J = 3.0, 4.8, 10.0)	5.92(ddd, J = 3.0, 4.8, 10.0)	5.40(br. d, J = 10.1)	5.92(ddd, J = 3.2, 4.8, 10.0)
H-C(9)	5.52(dd, J = 1.8, 10.0)	5.56(dd, J = 2.1, 10.1)	5.58(m)	5.54(ddd, J = 2.0, 4.2, 10.1)	5.55(dd, J = 1.2, 10.0)
H-C(10)	3.07(br. d, J = 8.8)	3.39(m)	3.20(dd, J = 1.2, 9.2)	3.15(br. s)	3.17(dd, J = 1.2, 8.5)
H-C(11)	4.96(d, J = 1.3)	4.47(t, J = 1.2)	5.05(d, J = 2.1)	1.47(s, 3H)	5.19(d, J = 2.0)
H-C(11)	4.61(d, J = 1.4)	4.38(t, J = 1.1)	4.82(d, J = 2.1)		4.91(d, J = 2.0)
H-C(12)	5.12(ddq, J = 9.1, 15.3, 1.6)	5.26(dd, j = 7.4, 15.5)	5.04(ddq, J = 9.2, 15.2, 1.7)	5.48(m)	5.15(ddq, J = 8.5, 15.5, 1.8)
H-C(13)	5.44(dq, J = 15.2, 6.2)	5.48(dq, J = 15.5, 6.4)	5.54(dq, J = 14.8, 6.6)	5.48(m)	5.47(dq, J = 15.5, 6.6)
H_3-C(14)	1.54(dd, J = 1.5, 6.4)	1.52(d, J = 6.2)	1.54(dd, J = 1.6, 6.5)	1.67(d, J = 3.7)	1.54(dd, J = 1.8, 6.4)
H_3-C(15)		3.69(s)	3.85(s)	3.72(s)	3.85(s)

[a] For all compounds, the numbering according to **1** is used.
[b] NH, 11.3 ppm (br. s, exchangeable with D_2O).
[c] Exchangeable with D_2O.
[d] Unsymmetrical.
[e] With D_2O sharper.

Table 11. ^{13}C-NMR Data $((D_6)DMSO)$ of the Spirostaphylotrichins P (28), O (29), N (30), T (31), and V (32)[a]

C-Atom	P (28)	O (29)	N (30)	V (32)
C(1)	174.8 (S)	169.2 (S)	167.0 (S)	167.1 (S)
C(3)	140.8 (S)	145.3 (S)	137.9 (S)	137.3 (S)
C(4)	200.9 (S)	65.1 (D)	195.5 (S)	196.1 (S)
C(5)	58.6 (S)	55.2 (S)	59.6 (S)	59.6 (S)
C(65)	70.9 (D)[b]	68.1 (D)[b]	70.4 (D)[b]	70.5 (D)[b]
C(7)	64.8 (D)[b]	64.7 (D)[b]	64.1 (D)[b]	64.5 (D)[b]
C(8)	129.5 (D)[c]	129.1 (D)[c]	130.2 (D)[c]	130.2 (D)[c]
C(9)	128.2 (D)[c]	128.0 (D)[c]	127.9 (D)[c]	127.7 (D)[c]
C(10)	45.1 (D)	39.3 (D)	44.0 (D)	44.7 (D)
C(11)	90.7 (T)	82.8 (T)	89.4 (T)	90.9 (T)
C(12)	128.5 (D)[c]	128.6 (D)[c]	128.8 (D)[d]	128.6 (D)[d]
C(13)	127.0 (D)[c]	128.0 (D)[c]	126.5 (D)[d]	126.4 (D)[d]
C(14)	17.5 (Q)	17.6 (Q)	17.4 (Q)	17.4 (Q)
C(15)		62.1 (Q)	62.7 (Q)	62.9 (Q)

[a] For all compounds, the numbering according to 1 is used.
[b,c,d] May be interchanged.

isolated from the mutant strain *P 649* is the presence of an hydroxyl function at C(7), in place of the oxo group, which was observed for compounds 1–7 and 19–26.

The assignment of the configuration for the spirostaphylotrichins N (30) and O (29) was derived from the results of a series of ^{1}H-NMR-NOE difference experiments, as well as on several considerations made on the basis of possible biogenetic correlations between these two substances and related spirostaphylotrichins. Some of these biogenetic considerations and correlations will be outlined in the following sections. Spirostaphylotrichin V (32), whose ^{1}H- and ^{13}C-NMR spectra are in good agreement with those of N, proved to be most likely the 6-epimer of 30 (27). Thus, N (30) and V (32) represent besides A (1) /B (2), C (3 or 4) /D (4 or 3), G (20 or 21) /H (21 or 20) an additional couple of diastereoisomers, differing from each other only in the relative configuration of C(6). This particular aspect of the spirostaphylotrichin production, namely the apparent parallel formation of epimeric compounds, was later subject of more detailed investigations (c.f. Chapter 4).

Compound P (28) differs from any other spirostaphylotrichin by the lack of the methoxy group usually attached to the nitrogen atom. Because of this structural particularity the hypothesis was put forward that P could be the earliest intermediate of the biosynthesis of spiro-

staphylotrichin A (1) having been isolated at present. Spirostaphylotrichin N (30) or its 6-epimer V (32) were considered as the most probable
successive biosynthetic products of P (28).

In order to verify these assumptions, it was tried to convert spirostaphylotrichin P into N (30) or V (32) by means of a cell-free reaction
with an enzyme extract obtained from ruptured cells of *S. coccosporum*
(27). This cell-free reaction was carried out at 28°, in a 0.1 M phosphate
buffer (pH 6.0), containing dithiothreitole (10 mM), EDTA (0.6 mM),
benzamidine hydrochloride (1 mM) and phenylmethanesulfonylfluoride
(0.6 mM), and in the presence of S-adenosyl-L-methionine (1 equivalent
SAM per each equivalent P (28)). The course of the reaction was followed
by HPLC analysis. Surprisingly it was observed, that P was completely
converted into spirostaphylotrichin O (29) under the conditions mentioned above, within only about 15 minutes[5]. This conversion took place
so rapidly, that it was not possible to detect any intermediates (c.f.
Scheme 6). Several experiments, performed with the aim to reduce the
reaction rates, turned out to be not very successful, but confirmed two
important aspects: (1) The conversion of P (28) into O (29) is selective.
Neither the formation of a possible epimer of O nor epimerization of
P were observed under the described reaction conditions. Therefore, it
seems justified to assume for P (28) the same absolute configuration at
C(6), C(7) and C(10) as for O (29). (2) This conversion, catalyzed by
enzymes present in the crude cell-free extract, takes place exclusively in
the presence of S-adenosyl-L-methionine. This fact confirms the results
obtained by feeding experiments with labelled L-methionine (Chapter
3.1. and 3.2), together with the hypothesis, that the 15-methyl group
originates from L-methionine, or rather from its activated form S-adenosyl-L-methionine.

4. Investigation of the Epimerization of
Spirostaphylotrichins

Having isolated from cultures of the wild type, as well as from those
of the mutant strains *P 84* and *P 649*, several pairs of epimers (A (1)/B (2),
C (3 or 4)/D (4 or 3), G (20 or 21) /H (21 or 20) and N (30)/V (32)), it was
interesting to investigate the possible reasons and mechanisms leading to

---

[5] Surprisingly this conversion did not require the addition of any particular coenzymes or
cofactors into the reaction mixture. These substances are probably present or are
regenerated in sufficient amounts in the crude extract.

the parallel formation of these compounds. In the course of the examination of the production spectrum in wild type cultures (cf. Chapter 2), we already have noted, that whereas spirostaphylotrichin A (1) was present in large amounts in fermentations in both kinds of production media (complex and minimal medium), its 6-epimer B (2) seemed to be produced almost exclusively in the minimal medium. We also had observed, that the culture broths of cultures in the minimal medium became more acidic as growth progressed (starting pH 6.0, final pH *ca.* 4.3); by contrast the pH in cultures grown in the complex medium remained more or less constant (pH *ca.* 6–7) during the whole fermentation. These observations suggested a possible relationship between the pH-value and the relative formation of A (1) and B (2). This assumption seemed to be supported by the observation, that the concentration of B (2) in cultures of *S. coccosporum* in the minimal medium diminished markedly when the pH of the culture broth was constantly maintained above 5.5, and that in favour of an increase of the accumulation of A (1). These interesting findings led us to investigate the pH-stabilities of spirostaphylotrichin A (1), B (2) and of the others epimeric compounds mentioned earlier. Small portions of these compounds (except for D, H and N, of which no sufficient amounts were available) were suspended separately in two buffers with pH 7.0 and 4.5 and these solutions were stirred at 28° for 24 hours. Samples of these suspensions were analyzed by HPLC in order to ascertain possible spontaneous epimerizations. The results, as summarized in Table 12, demonstrate that every compound tested is partially converted into its 6-epimer, depending on the pH-value of the solution.

Since these different reversible epimerizations take place quite rapidly and under normal fermentation conditions, it is possible that only one compound of every epimeric pair is a real product or intermediate of the spirostaphylotrichin biosynthesis, while the other one is an artefact.

Table 12. *Ratio of [Epimer 1] : [Epimer 2] in Different Buffer Solutions*

Epimer 1	[Epimer 1]:[Epimer 2] in buffer pH 7.0 (24 h)	[Epimer 1]:[Epimer 2] in buffer pH 4.5 (24 h)	Epimerization
A (1)	88.8:11.2	85.3:14.7	A (1) → B (2)
B (2)	66.5:33.5	81.0:19.0	B (2) → A (1)
C (3 or 4)	65.7:34.3[a]	33.1:66.9	C (3/4) → D (4/3)
G (20 or 21)	84.0:16.0	93.6:6.4	G (20/21) → H (21/20)
V (32)	~ 100:0	48.1:51.9	V (32) → N (30)

[a] Sample was stirred only 20 minutes in the buffer solution. Longer periods of times caused the decomposition of C.

These spontaneous isomerization might be explained by the mechanism as illustrated in the case of the spirostaphylotrichins N (**30**) and
V (**32**) in Scheme 4. Spirostaphylotrichin N is converted in a retro-aldol
reaction into the hypothetical anionic intermediate **33**, in which the
carbon atoms No. 5 and 6 (sp^2-hybridized) are achiral. **33** in turn cyclises
to give V (**32**) or again N (**30**). This reversible epimerization would
require, in the form shown in Scheme 4, a specific conversion course
related to the configuration of C (5).

In the case of the spirostaphylotrichins A (**1**), B (**2**), C (**3** or **4**), D (**4** or
3), G (**20** or **21**) and H (**21** or **20**), possessing a carbonyl group at position
No. 7, the respective epimerizations could occur also following an additional mechanistic pathway, which implies, as shown in Scheme 5, the
formation of ene-diole intermediates by keto-enol tautomerie. Actually,
we were able to ascertain the presence of ene-diols in slightly acidic
aqueous solutions of the compounds A, B, C and G (ene-diol detection
through complexation with titan (IV) reagents (*29*)).

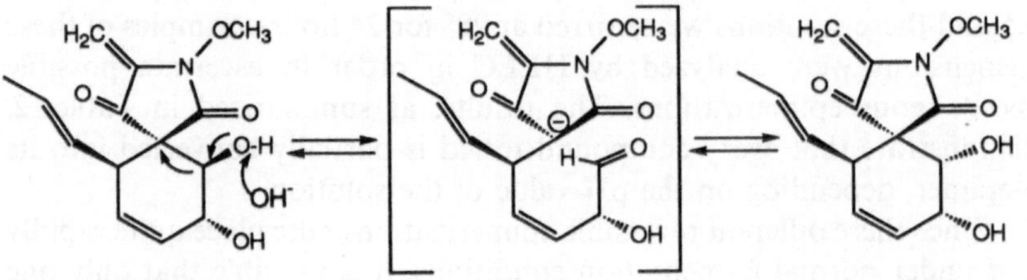

Scheme 4

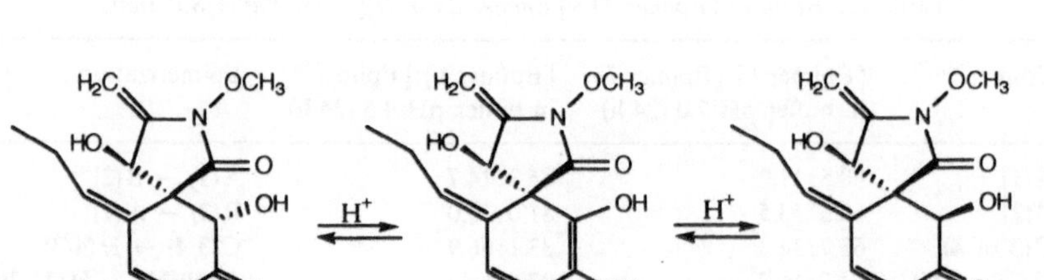

5. Discussion of the Results Regarding the Biosynthesis of the Spirostaphylotrichins

With the goal in mind to facilitate the interpretation of possible relationships between the various compounds isolated from *S. cocco-sporum* (wild type) and the two mutants, we divided the spirostaphylo-trichins into two categories: (1) Spirostaphylotrichins representing prob-able, real products or intermediates of the spirostaphylotrichin biosynthe-sis and (2) compounds, which are formed from the spirostaphylotrichins belonging to category 1 via spontaneous epimerizations or other undesir-able reactions (*e.g.* hydrolysis) during the isolation procedure, or that result from enzymatic secondary reactions.

The compounds of this second group of spirostaphylotrichins, among which figure Q (**5**), R (**6**) and presumably also F (**7**), E (**19**), I (**22**), L (**24**), M (**25**) S (**26**) and T (**31**), were then excluded in order to simplify the situation from the considerations reported further below.

A structural characteristic common to all metabolites which are considered to be real biosynthetic products or intermediates (*e.g.* A (**1**), P (**28**), O (**29**) and V (**32**)) is the absolute configuration of C(6). Spiro-staphylotrichin P (**28**) represents the earliest biosynthetic intermediate having been isolated. All others intermediates or products are biogeneti-cally derived from this compound.

A probable sequence of the terminal steps of the spirostaphylotrichin biosynthesis, starting from P (**28**), is illustrated in Scheme 6. The conver-sion of P into V (**32**) involves the hydroxylation of the lactam nitrogen atom (Step 1), leading to the hypothetical intermediate **35**, followed by the methylation of the N–OH group by S-adenosyl-L-methionine (Step 2). From V (**32**) the biosynthetic pathway divide. Following the main route, V (**32**) is reduced to form O (**29**) (Step 3). In the latter the 7-hydroxy group is oxidized (Step 4) to yield the intermediate **36**[6], which is the 6-epimer of K (**23**). Subsequent isomerization of the double bond in the side chain leads to spirostaphylotrichin A (**1**) (Step 5). In a parallel pathway smaller amounts of spirostaphylotrichin V (**32**) are converted to compound **20** (G or H; Step 4') by selective oxidation of the 7-hydroxy group. By isomerization of the double bond in the side chain compound **3** is obtained, which corresponds either to spirostaphylotrichin C or D (Step 5'). Such parallel biosynthetic pathways represent a frequent phenomenon of the secondary metabolism (*30*). They are caused by the

---

[6] Compound **36**, postulated here as biosynthetic intermediate and direct precursor of A (**1**), probably figure among the minor metabolites, having not yet been isolated, present in culture extracts of the mutant strain *P 84*.

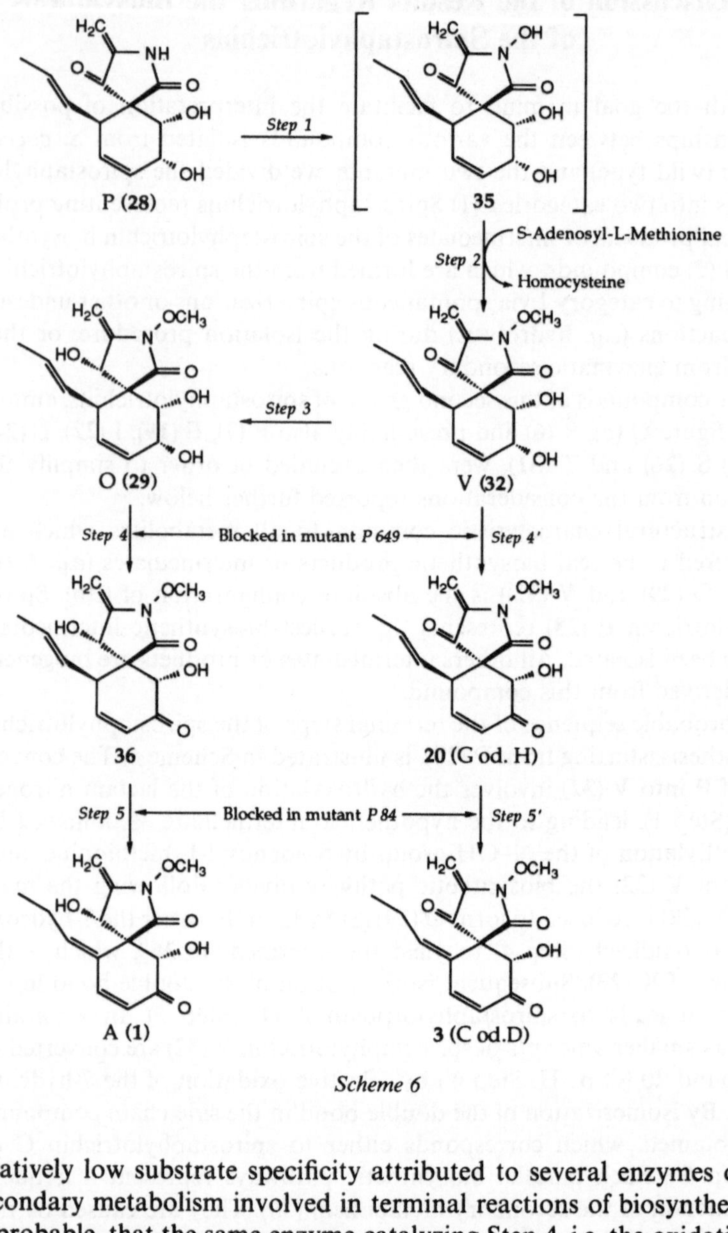

Scheme 6

relatively low substrate specificity attributed to several enzymes of the secondary metabolism involved in terminal reactions of biosynthesis. It is probable, that the same enzyme catalyzing Step 4, i.e. the oxidation of the 7-OH function in spirostaphylotrichin O (29), converts also V (32) into 20 (Step 4'). In analogy, the enzyme catalyzing the last reaction of the spirostaphylotrichin biosynthesis, namely the shifting of the side chain double bond, converts compound **36** into A (**1**), as well as **20** (G or H) into **3** (C or D).

Concerning the sequence of the biosynthetic steps preceeding the reactions just mentioned, one can only formulate a vague hypothesis. As already illustrated in detail in Chapter 3, the carbon skeleton of spiro-staphylotrichin A (**1**) (and of all spirostaphylotrichins biogenetically

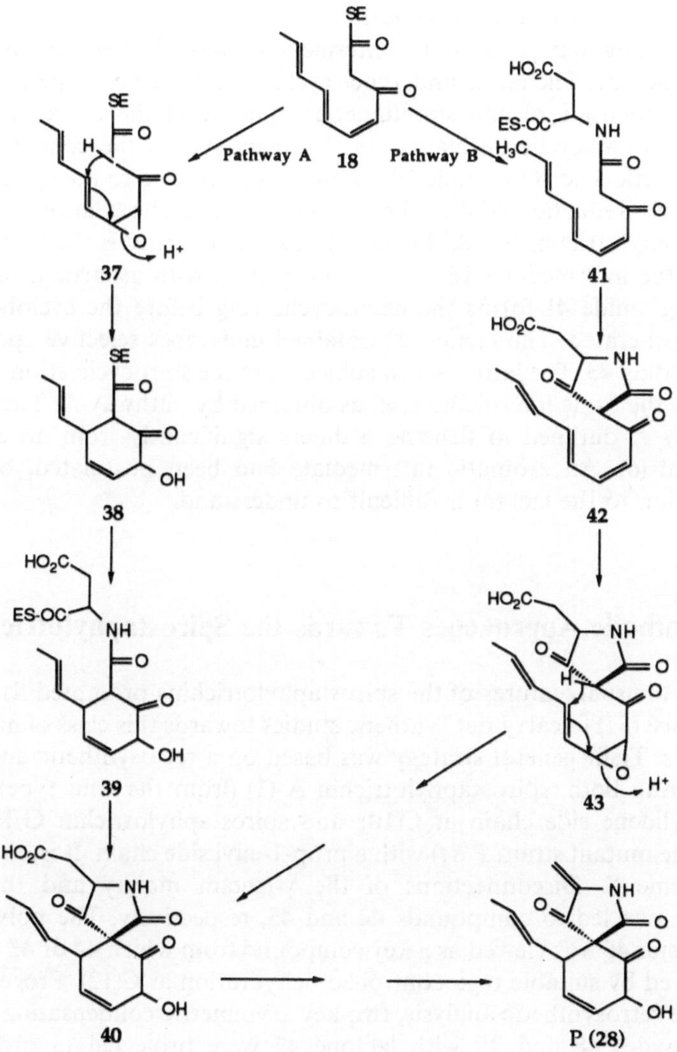

Scheme 7

related to **1**) arises from the condensation of a pentaketide with a C_4-dicarboxylic acid biogenetically linked to the citric acid cycle. We suppose it to be L-aspartic acid. In Section 3.2 we also have presented a hypothetical pathway for the biosynthesis of the postulated pentaketide intermediate **18** (Scheme 2). Starting from the assumption that the pentaketide which is involved in the spirostaphylotrichin biosynthesis, possesses the structure **18**, one may propose for the sequence of the subsequent biosynthetic steps one of the two possibilities (pathway A or B) which are outlined in Scheme 7.

According to pathway A, the intermediate **18** is selectively oxidized to the epoxide **37**. The latter undergoes cyclization by a S_N2'-type reaction (aldol condensation) and simultaneous opening of the oxiran ring to form the monocyclic derivative **38**. The next step is the condensation with aspartic acid. The amide **39** obtained is converted to the spirocyclic lactam **40**. Reduction of the 6-keto group and decarboxylation leads to spirostaphylotrichin P (**28**). In the alternative pathway B the first reaction of the intermediate **18** is the condensation with aspartic acid. The resulting amide **41** forms the heterocyclic ring before the cyclohexane ring is generated. The lactam **42** obtained undergoes selective epoxidation yielding **43**. The latter is now subjected to the spirocyclization which leads to the same intermediate **40** as obtained by pathway A. The main pathway A outlined in Scheme 6 differs significantly from an earlier proposal (*6*). An aromatic intermediate had been postulated, but its cyclization to the lactam is difficult to understand.

6. Synthetic Approaches Towards the Spirostaphylotrichins

The unusual features of the spirostaphylotrichins prompted STEINER and TAMM (*31*) to carry out synthetic studies towards this class of natural products. Their general strategy was based on a retrosynthetic analysis considering both, spirostaphylotrichin A (**1**) (from the wild type) with a propylidene side chain at C(10), and spirostaphylotrichin G/H (**20**) (from the mutant strain *P 84*) with a prop-1-enyl side chain. It is outlined in Scheme 8. Disconnections of the γ-lactam moiety and the cyclohexenone led to compounds **44** and **45**, respectively. The polyfunctional ester **46** was viewed as a key compound from which **44** or **45** could be derived by suitable regiocontrolled dehydration at C(12). Proceeding with the retrosynthetic analysis, two key asymmetric condensation steps of aldehydes **48** and **49** with lactone **47** were projected in order to generate the appropriate configuration at the quaternary centre **46**

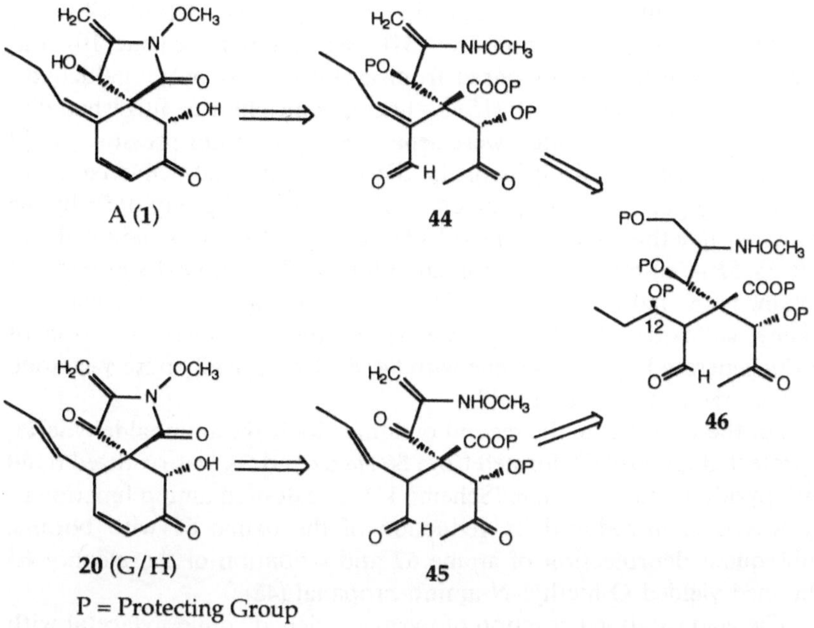

A (1)   44

20 (G/H)   45

46

P = Protecting Group

Scheme 8

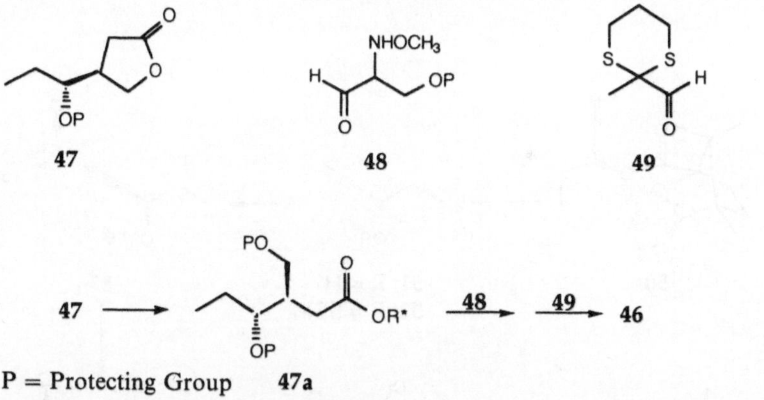

47   48   49

47 ⟶   47a   48   49   46

P = Protecting Group   47a

Scheme 9

(Scheme 9). Prior to these connections the γ-butyrolactone **47** should be hydrolysed and esterified with a chiral auxiliary to give **47a**. The required configuration of the quaternary centre of **46** should then be established in **47a** by asymmetric aldol condensations with **48** and **49**.

For the synthesis of the optically pure γ-butyrolactone **47** enantiomerically pure (3R)-pentin-3-ol (**51**) was required (Scheme 10). Racemic pentinol **50** was prepared from acetylene and propionaldehyde. Reaction of racemic **51** with (1S)-(−)-camphanic chloride furnished diastereoisomeric esters which were separated by medium pressure liquid chromatography. Hydrolysis of the diastereoisomer **50** delivered enantiomerically pure **51**. After protection of the hydroxyl group of **51** by the trimethylsilylethoxymethoxyl (SEM) group, **52** was coupled to (1R,2S,5R)-(−)-methyl-(S)-4-toluenesulfinate. The alkynyl sulfoxide **53** obtained was reduced with LiAlH$_4$ at low temperature to furnish (E)-alkenyl sulfoxide **54**. The asymmetric additive PUMMERER reaction of *in situ* generated dichloroketene with **54** yielded optically pure γ-lactone **55**, the precursor of lactone **47**.

For the synthesis of the second building block, the aminoaldehyde **48**, diprotected glycerol **59**, formed from **56** *via* **57** and **58**, was oxidized to **60** with pyridinium dichromate (Scheme 11). The desired amino functionality was then introduced by reduction of the oxime **61** with borane. Subsequent deprotection of amine **62** and oxidation of the alcohol **63** obtained yielded O-methyl-N-amino-propanal (**48**).

The acid catalysed reaction of pyruvic aldehyde dimethylacetal with 1,3-propanedithiol generated aldehyde **49**, the third building block.

The synthesis of lactone **47** and its connection with the building blocks **48** and **49** have not been reported yet.

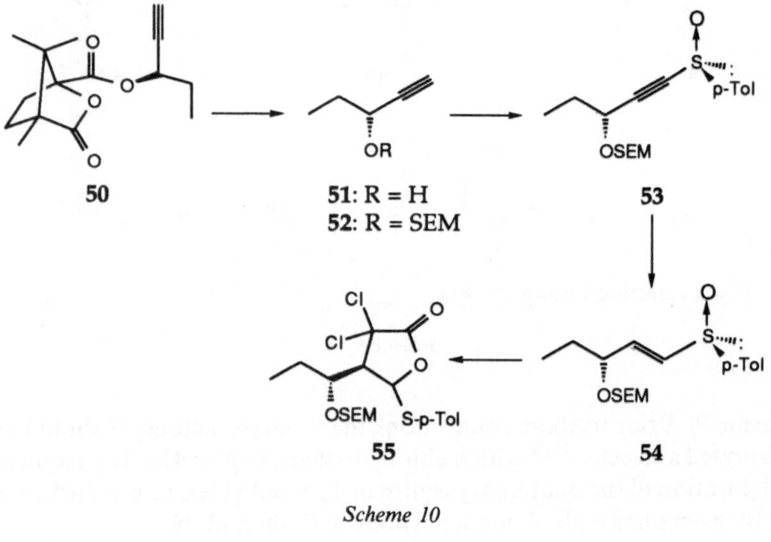

50

51: R = H
52: R = SEM

53

55

54

Scheme 10

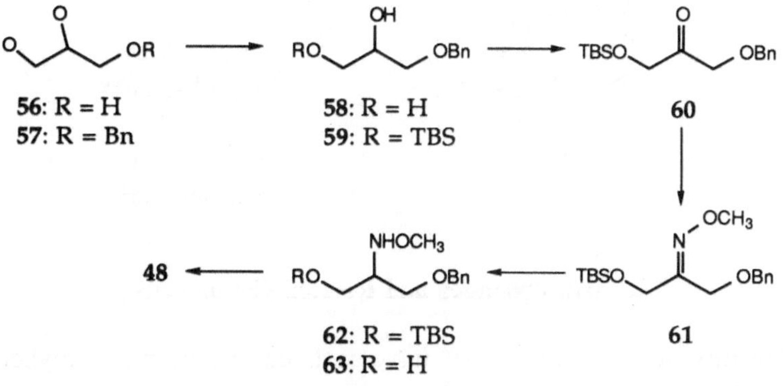

Scheme 11

7. Related Fungal Metabolites: Triticones, Arthropsolides and Other Compounds

7.1. Triticones

Exploring the possibility of characterizing novel phytotoxins from microorganisms that attack weeds, with the aim to use these structures as leads for new herbicides, SUGAWARA et al. isolated from cultures of the plant pathogenic fungus *Drechslera tritici-repentis* several compounds, named triticones A, B, C, D, E, and F *(7, 8)*. All six metabolites were present in submerged cultures of *D. tritici-repentis* in relatively small concentrations, at levels of 6–16 mg per litre culture broth.

The triticones A and B proved to be the compounds possessing the highest biological (phytotoxic) activity. The structural elucidation of the triticones, based on the data of spectroscopic measurements and X-ray analysis, revealed the close chemical similarity between these compounds and the spirostaphylotrichins. In fact, the triticones A, B, C and D possess the same constitution and the same relative configuration having been attributed to the spirostaphylotrichins 4 (C/D), 3 (D/C), A (1) and B (2) respectively. However, since the absolute configuration of the triticones has not been determined, it is still unclear whether these substances are identical with the spirostaphylotrichins mentioned above or correspond contrarily to their enantiomers. Concerning the triticones E and F, the authors have suggested for this two compounds the structures **64** and **65** (*cf.* spirostaphylotrichin R *(7)*), whose correctness however, with regard to the absolute configuration, has not been confirmed experimentally.

Triticone E (64)   Triticone F (65)

7.2. Arthropsolides and Related Compounds

Recently AYER et al. (9, 10) have isolated twelve new polyketide metabolites from submerged cultures of the fungus *Arthropsis truncata* (*Fungi imperfecti, Hyphomycetes*) and determined their structures by a combination of chemical and spectroscopic methods. Four of these compounds possess an unusual 2-oxaspiro[4, 5]decane skeleton, which shows a striking analogy to the 2-azaspiro[4, 5]decane system of the spirostaphylotrichins and triticones. These closely related metabolites are arthropsolide A (66), B (67), C (68) and D (69). The structures of 66 has been confirmed by an X-ray crystallographic study of the di-O-benzoyl derivative. The absolute configuration of arthropsolide A (66), as well as those of the other arthropsolides (67–69), has been predicted by application of the HARADA-NAKANISHI dibenzoate rule (32) to the CD-spectrum of the tri-O-benzoyl derivative of arthropsolide D (69).

Besides the four spirocyclic γ-lactones two cyclohexenone derivatives, arthropsadiol A (70) and B (71), as well as two cyclohexenols, arthropsatriol A (72) and B (73) have been isolated. Of the remaining metabolites obtained structure 74 assigned to cycloarthropsadiol C is remarkable because of the combination of an acetal and hemiacetal in a bridged system. The structure of cycloarthropsone (75) is less complex being a 7-hydroxy-4(E)-propenyl-3(2H)-benzofuranone derivative. Finally the arthropsatriols C (76) and D (77) are epimers, which were only obtained as inseparable mixture.

Regarding the biosynthesis of the metabolites, the incorporation of [1-^{13}C]- and [1,2-^{13}C]acetate into arthropsadiol A (70), cycloarthropsone (75) and arthropsatriol C (76) has been examined (10). It was concluded that a hexaketide 78, formed from one acetate and five malonate units, is an early intermediate of the biosynthetic pathway of 70 (Scheme 12). In the course of the subsequent conversion from 78 to 70 via the intermediates 79, 80, 81 and 82, the last step from 82 to 70 requires the introduction of a C$_1$-unit. It probably is provided by S-adenosyl-L-methionine.

Arthropsolide A (66)   Arthropsolide B (67)   Arthropsolide C (68)

Arthropsolide D (69)   Arthropsadiol A (70)   Arthropsadiol B (71)

Arthropsatriol A (72)   Arthropsatriol B (73)   Cycloarthropsadiol C (74)

Cycloarthropsone (75)   Arthropsatriol C (76)   Arthropsatriol D (77)

No incorporation experiments have been performed for the elucidation of the biosynthesis of arthropsolides 66–69. However it is reasonable to assume, that the biosynthesis of arthropsolide A (66) and B (67), which are very similar to the spirostaphylotrichines K (23), C (3/ 4) and D (4/3), starts also with the formation of a pentaketide, which condenses in a later step with one equivalent of malic acid in place of one equivalent of aspartic acid as it is the case in the spirostaphylotrichin series. In analogy

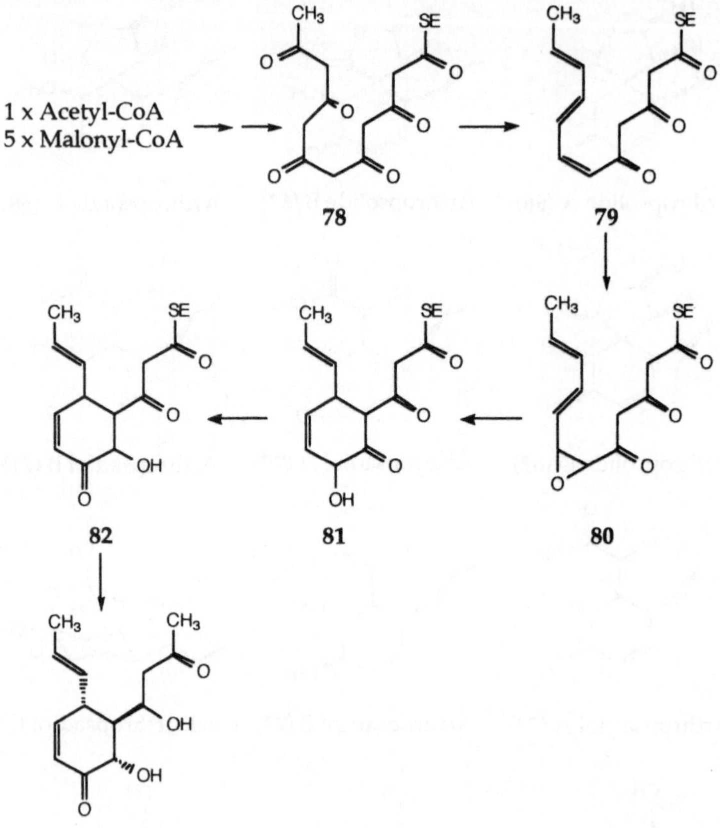

1 x Acetyl-CoA
5 x Malonyl-CoA

Scheme 12

to the biosynthetic pathway A, which we have proposed for the spiro-staphylotrichins (cf. Scheme 7), a reaction sequence is feasable for the arthropsolides A (66) and B (67), which at the same time would provide a plausible biogenetic relationship between arthropsatriol A (72) and arthropsadiol A (70) (Scheme 13).

Starting from one acetyl- and four malonyl-coenzyme A units the pentaketide 83 is formed. Selective epoxidation leads to the epoxide 84. The latter cyclizes to compound 85 by a S_N2'-like reaction accompanied by opening of the oxiran ring. By condensation with one equivalent of malic acid, 85 is converted to the intermediate 86. The latter undergoes spirocyclization to form the γ-lactone 87. A series of further transforma-

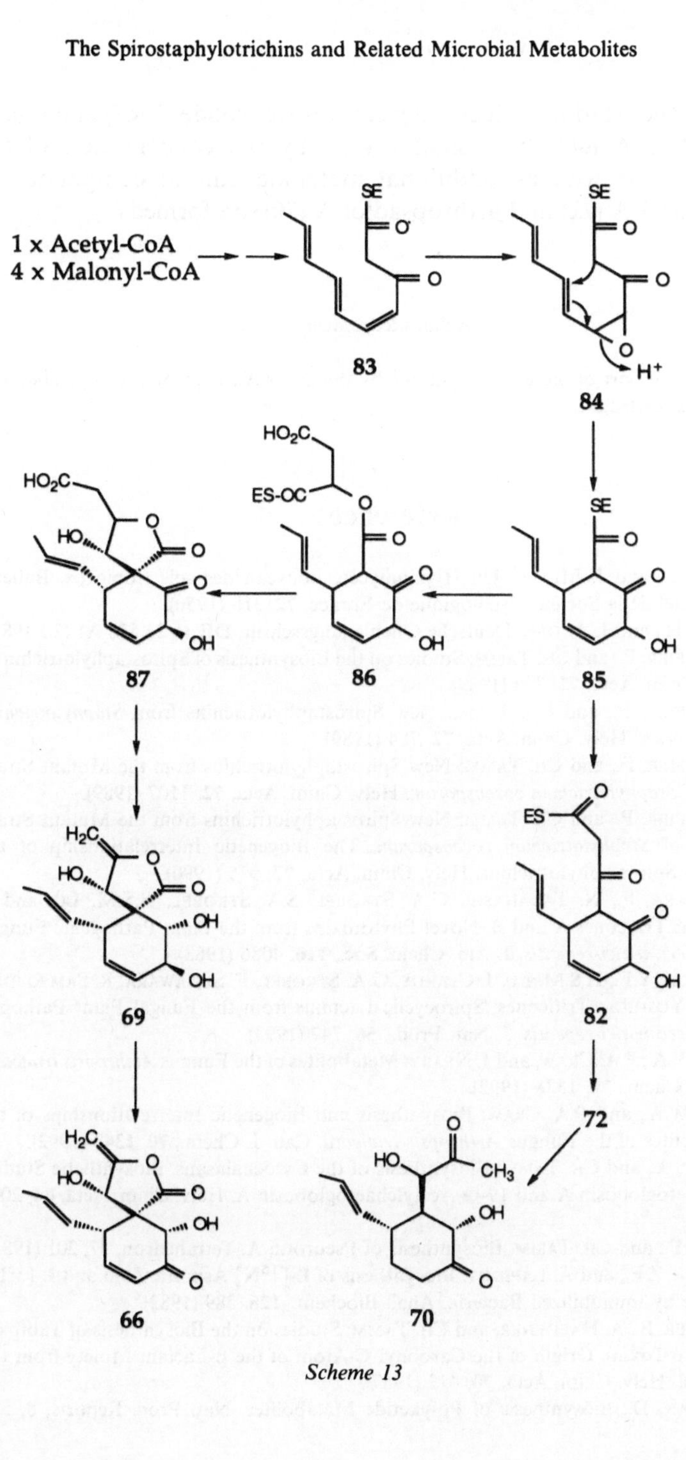

Scheme 13

tions (decarboxylation, reduction) yields arthropsolide D (69) and finally arthropsolide A (66). In a similar way, by the condensation of the intermediate 85 with an additional malonate unit to compound 82, arthropsatriol A (72) and arthropsadiol A (70) are formed.

Acknowledgement

Financial support of these investigations by the *Swiss National Science Foundation* is gratefully acknowledged.

References

1. NICOT, J., and J. MEYER: Un Hyphomycète nouveau des sols tropicaux. Bulletin trimestriel de la Société mycologique de France, **72**, 318 (1956).
2. PETER, H., and J. AUDEN: Deutsche Offenlegungsschrift, DE 35 22 578 A1 (2.1.1986).
3. SANDMEIER, P., and CH. TAMM: Studies on the Biosynthesis of Spirostaphylotrichin A. Helv. Chim. Acta, **72**, 774 (1989).
4. SANDMEIER, P., and CH. TAMM: New Spirostaphylotrichins from *Staphylotrichum coccosporum*. Helv. Chim. Acta, **72**, 784 (1989).
5. SANDMEIER, P., and CH. TAMM: New Spirostaphylotrichins from the Mutant Strain *P 84* of *Staphylotrichum coccosporum*. Helv. Chim. Acta, **72**, 1107 (1989).
6. SANDMEIER, P., and CH. TAMM: New Spirostaphylotrichins from the Mutant Strain *P 649* of *Staphylotrichum coccosporum*: The Biogenetic Interrelationship of the Known Spirostaphylotrichins. Helv. Chim. Acta, **73**, 975 (1990).
7. SUGAWARA, F., N. TAKAHASHI, G.A. STROBEL, S.A. STROBEL, H.S.M. LU, and J. CLARDY: Triticones A and B, Novel Phytotoxins from the Plant Pathogenic Fungus *Drechslera tritici-repentis*. J. Am. Chem. Soc., **110**, 4086 (1988).
8. HALLOCK, Y.F., H.S.M. LU, J. CLARDY, G.A. STROBEL, F. SUGAWARA, R. SAMSOEDIN, and S. YOSHIDA: Triticones, Spirocyclic Lactams from the Fungal Plant Pathogen *Drechslera tritici-repentis*. J. Nat. Prod., **56**, 747 (1993).
9. AYER, W.A., P.A. CRAW, and J. NEARY: Metabolites of the Fungus *Arthropsis truncata*. Can. J. Chem., **70**, 1338 (1992).
10. AYER, W.A., and P.A. CRAW: Biosynthesis and Biogenetic Interrelationships of the Metabolites of the Fungus *Arthropsis truncata*. Can. J. Chem., **70**, 1348 (1992).
11. PROBST, A., and CH. TAMM: Biosynthesis of the Cytochalasans. Biosynthetic Studies on Chaetoglobosin A and 19-O-Acetylchaetoglobosin A. Helv. Chim. Acta, **64**, 2065 (1981).
12. MOHR, P., and CH. TAMM: Biosynthesis of Pseurotin A. Tetrahedron, **37**, 201 (1981).
13. KAHANA, Z.E., and A. LAPIDOT: Biosynthesis of L-[^{15}N] Aspartic Acid and L-[^{15}N] Alanine by Immobilized Bacteria. Anal. Biochem., **126**, 389 (1982).
14. MUELLER, B., A. HAEDENER, and CH. TAMM: Studies on the Biosynthesis of Tabtoxin (Wildfire Toxin). Origin of the Carbonyl C-Atom of the β-Lactam Moiety from the C_1-Pool. Helv. Chim. Acta, **70**, 412 (1987).
15. O'HAGAN, D.: Biosynthesis of Polyketide Metabolites. Nat. Prod. Reports, **8**, 573 (1991).

16. HOPWOOD, D.A., and C. KHOSLA: Genes for Polyketide Secondary Metabolic Pathways in Microorganisms and Plants. In: Secondary Metabolites: Their Function and Evolution (Ciba Foundation Symposium, 171), p. 88. Chichester-New York: Wiley. 1992.

17. CANE, D.E., and C.C. YANG: Macrolide Biosynthesis. Intact Incorporation of a Chain-Elongation Intermediate into Erythromycin. J. Am. Chem. Soc., 109, 1255 (1987).

18. CANE, D.E., P.C. PRABHAKARAN, W. TAN, and W.R. OTT: Macrolide Biosynthesis. Mechanism of Polyketide Chain Elongation. Tetrahedron Lett., 32, 5457 (1991).

19. DONADIO, S., M.J. STAVER, J.B. MCALPINE, S.J. SWANSON, and L. KATZ: Modular Organization of Genes Required for Complex Polyketide Biosynthesis. Science, 252, 675 (1991).

20. CANE, D.E., R.H. LAMBALOT, P.C. PRABHAKARAN, and W.R. OTT: Incorporation of Polyketide Chain Elongation Intermediates into Methymycin. J. Am. Chem. Soc., 115, 522 (1993).

21. YUE, S., J.S. DUNCAN, Y. YAMAMOTO, and C.R. HUTCHINSON: Macrolide Biosynthesis. Tylactone Formation Involves the Processive Addition of Three Carbon Units. J. Am. Chem. Soc., 109, 1253 (1987).

22. CANE, D.E., W. TAN, and W.R. OTT: Nargenin Biosynthesis. Incorporation of Polyketide Chain Elongation Intermediates and Support for a Proposed Intermolecular Diels-Alder Cyclization. J. Am. Chem. Soc., 115, 527 (1993).

23. PATZELT, H., and J.A. ROBINSON: Biosynthesis of the Polyether Antibiotic Monensin A: Incorporation of a Polyketide Chain Elongation Intermediate. J. Chem. Soc., Chem. Commun., 16, 1258 (1993).

24. HAILES, H.C., C.M. JACKSON, P.F. LEADLAY, S.V. LEY, and J. STAUNTON: Biosynthesis of Tetronasin, Part 1: Introduction and Investigation of the Diketide and Triketide Intermediates Bound to the Polyketide Synthase. Tetrahedron Lett., 35, 307 (1994).

25. HAILES, H.C., H. SANDEEP, F. LEADLAY, I.C. LENNON, S.V. LEY, and J. STAUNTON: Biosynthesis of Tetronasin, Part 2: Identification of the Tetraketide Intermediate Attached to the Polyketide Synthase. Tetrahedron Lett., 35, 311 (1994).

26. HAILES, H.C., H. SANDEEP, F. LEADLAY, I.C. LENNON, S.V. LEY, and J. STAUNTON: Biosynthesis of Tetronasin, Part 3: Preparation of Deuterium Labelled Tri- and Tetraketides as Putative Biosynthetic Precursor of Tetronasin. Tetrahedron Lett., 35, 315 (1994).

27. WALSER-VOLKEN, P.: Ph.D. Thesis. Basel: 1993.

28. ALLINGER, N.L., and U. BURKERT: Molecular Mechanics (ACS Monograph, 171). Washington, DC: American Chemical Society. 1982.

29. HESSE, G.: In: Methoden der organischen Chemie (HOUBEN-WEYL), Vol. VI/1d, 4th Ed. (E. MÜLLER, O. BAYER, eds.), p. 224. Stuttgart: Thieme. 1978.

30. LUCKNER, M.: Secondary Metabolism in Microorganisms, Plants and Animals, 3rd Ed. Berlin-Heidelberg-New York: Springer. 1990.

31. STEINER, O., and CH. TAMM: Synthetic Studies Towards Spirostaphylotrichins: Synthesis of Building Blocks. Tetrahedron Lett., 34, 6729 (1993).

32. HARADA, N., and K. NAKANISHI: A Method for Determining the Chiralities of Optically Active Glycols. J. Am. Chem. Soc., 91, 3989 (1969).

(Received December 13, 1994)

Author Index

Page numbers printed in *italics* refer to References

Aguiar, Z.E. *122*
Aguilar, M.A. *118*
Ahmed, V.U. *114*
Alcorn, J.B. *116*
Alicia, D. *116*
Allinger, N.L. *165*
Alonso, G. *118*
Alvarenga, N.L. *123*
Alves, R.J. *122*
Amezcua, B. *116*
Angeletti, P.V. 99, 100, *115*
Asfora, J.J. *117*
Atta-ur-Rahman *114*
Auden, J. 126, *164*
Avendano, S. *116*
Ayengar, K.N.N. *122*
Ayer, W.A. 127, 160, *164*

Balasubramaniam, S. *117, 120, 122, 123*
Bansal, R.K. *120*
Bansiddhi, J. *121*
Barreto, V.M. *115*
Bavovada, R. *115, 121*
Baxter, R.L. *115*
Bazzocchi, I.L. *118, 122, 123*
Bazzocchi, L.L. *119*
Beale, M.H. *120*
Berry, D.E. *116*
Bhargava, P.N. 4, *115*
Bhatnagar, S.S. 4, 37, 38, 74, 96, 97, *115, 120*
Blasko, G. *115*
Boada, J. *118, 119*
Braeden, O.J. *115*
Bröckelmann, M. *123*

Brown, P.M. *115*
Brüning, R. 2, 19, 95, *115*
Budzikiewicz, H. 71, *121*
Burkert, U. *165*
Bye, R. *115*

Calvalcanti, M. da S.B. *117*
Calzada, F. *115*
Campanelli, A.R. *115*
Campbell, J. *116*
Cane, D.E. 142, *165*
Cano, G. *116, 118*
Cardenas, J. *116*
Carlisle, C.H. 74, *115*
Chalmers, W.T. *120*
Chaudhuri, S.K. 59, *121*
Chou, T.Q. 3, *115*
Clardy, J. *164*
Cooke, R.G. 39, 40, *115*
Cordell, G.A. 7, 49, 59, 60, *115, 121*
Cortias, C.T. *117*
Courtney, J.L. *115*
Craw, P.A. *164*
Crespo, A. *123*
Crombie, L. *115*

D'Alagni, M. 101, *115*
D'Albuquerque, I.L. *115–118, 121*
Darias, V. *118*
De A. Lima, D. *118*
De Barros Coelho, J.S. *115, 116–118*
Dechatiwongse, T. *121*
De las Heras, F.G. *118*
Delle Monache, F. *115, 116, 121, 122*
Del Rio, F. *115*

De Luca, C. *116*
De Mello, J.F. *115, 116, 121*
De M. Souza, M.A. *118*
De Santana, C.F. *121*
De Souza, J.R. *122*
Dev, S. *116*
De V. Pinto, K. *117*
Dhanabalasingham, B. *116, 122*
Dias, M.N. 59, *116, 122*
Divekar, P.V. 4, 37, 96, 97, *115*
Dominguez, X.A. *116, 118, 119*
Dominguez Jr., X.A. *116*
Donadio, S. *165*
Dorta, R.L. *118, 119*
Duncan, J.S. *165*
Dutta, N.L. *115*

Edwards, J.M. *116*
Ehrenberg, M. 74, *115*
Esquivel, B. *116*
Estrada, R. *116*

Fang, S.D. *116*
Farnsworth, N.R. 59, *121*
Fayos, J. *118*
Fernandes, F. 4, *120*
Fernando, C. *119*
Fernando, H.C. *116, 122*
Ferreira de Santana, C. 99, 100, *117*
Ferro, B.E. *118*
Ferro, E.A. *118, 122*
Fieser, L.F. 37, *117*
Filho, J.L. *121*
Fraga, B.M. *118*
Francisco, C.G. *118*
Franco, R. *116*
Freire, R. *118*
Fröde, R. *123*

Gamlath, C.B. *117, 122*
Garcia, S. *116*
Giglio, E. *115*
Gisvold, O. 3, 4, 35, 37, 38, 77, 79, *117*
Goncalves de Lima, C. *118*
Goncalves de Lima, O. 97, *115–118, 121*
Gonzalez, A.G. 75, 97–100, *118, 119, 121, 123*
Gonzalez, C.M. *118, 122*
Gonzalez, P. *118*
Goodlett, V.W. *123*
Goto, M. *123*

Govindachari, T.R. 36, 59, *121*
Grant, P.K. 39, 41, 80, 82, 83, *119*
Gullo, V.P. *121*
Gunaherath, G.M.K.B. *119*
Gunaherath, K.B. *117, 120*
Gunatilaka, A.A.L. *116, 117, 119, 120, 122*
Gunawardena, V. *120*
Gupta, A.S. *116*
Gutierrez, A.M. *119*
Gutierrez, A.N. *118*
Gutierrez, J. *118*
Gutierrez-Navarro, A.M. *118, 121*

Haedener, A. *164*
Hailes, H.C. *165*
Haller, H.L. 4, 37, 79, *120, 122*
Hallock, Y.F. *164*
Ham, P.J. 74, *120*
Harada, N. 160, *165*
Harada, R. *120*
Hecht, S.M. *116*
Hegnauer, R. 19, *120*
Heringer, E.P. *118*
Hernandez, R. *118*
Herrera, H.R. *122*
Hesse, G. *165*
Hewitt, G.M. *120*
Heywood, V.H. 19, *120*
Hirata, Y. *121*
Hopwood, D.A. *165*
Horii, S. *123*
Hutchinson, C.R. 142, *165*

Iitaka, Y. *120, 123*
Ikuta, H. *120*
Ireland, R.E. *120*
Itai, A. *120*
Itokawa, H. 108, 109, *120, 123*

Jackson, C.M. *165*
Jacoli, G.G. *120*
Jain, M.K. *120*
Jannotti, N.K. *122*
Jardim, M.L. *121*
Jayasena, K. *120*
Jimenez, A. *122*
Jimenez, I.A. *123*
Jimenez, J.S. *118, 119, 122*
Johnson, A.W. 39–42, 77, 80–84, *119, 120*

Jones, R.N.   37, *117*
Joshi, K.C.   *120*
Juby, P.F.   *119, 120*

Kahana, Z.E.   *164*
Kakisawa, H.   *120, 121*
Kamat, V.N.   4, *120*
Karunaratne, V.   *116, 122*
Katz, L.   *165*
Khosla, C.   *165*
Kikuchi, T.   *116, 117, 119, 122*
King, T.J.   *115, 119, 120*
Kobayashi, S.   *120*
Krishnamoorthy, V.   *115, 120*
Krishnan, V.   *120*
Kulkarni, A.B.   4, 37–40, 43, 74, 82, *120,*
   122
Kumar, V.   *116, 120, 123*
Kun-Anake, A.   *121*
Kurihara, T.   *120*
Kutney, J.P.   24, 104, *120*

Lacerda, A.L.   *117*
Lacet, Y.   *121*
Lambalot, R.H.   *165*
Lapidot, A.   *164*
Leadlay, F.   *165*
Leadlay, P.F.   *165*
Lennon, I.C.   *165*
Lewis, R.J.   *121*
Ley, S.V.   *165*
Likhitwitayawuid, K.   *121*
Lin, L.-Z.   *121*
Linares, E.   *115*
Lopez, I.   *118*
Lopez, R.   *115*
Lu, H.S.M.   *164*
Luckner, M.   *165*
Luis, J.G.   *118, 119, 121, 122*
Lynn, D.G.   *116*
Lynn, W.S.   *116*

Maciel, G.M.   *117, 118*
Mahato, S.B.   *121*
Marini-Bettolo, G.B.   49, 83, 93, 99, 100,
   104, *115, 116, 118, 121, 122*
Marta, M.   *121*
Martin, J.D.   46, *121*
Martinez, M.   *118*
Martinod, P.   *121*

Martins, D.G.   *117*
Marumoto, R.   *123*
Mata, R.   *115*
McAlpine, J.B.   *165*
Mei, P.F.   3, *115*
Melo, A.M.   *121*
Mendoza, J.J.   *119*
Meyer, J.   *164*
Mhaskar, V.V.   *122*
Miura, L.   *121*
Miyake, A.   *123*
Mohr, P.   *164*
Moir, M.   *115, 120*
Monache, F.D.   *118*
Monache, G.D.   *118*
Moreira, I.C.   *117*
Morita, H.   *120, 123*
Moujir, L.M.   *118, 119, 121*
Mueller, B.   *164*
Munoz, O.M.   *123*
Musya, M.   *120*

Nakanishi, K.   4, 36, 38, 39, 41, 42, 59, 71,
   77, 80, 84, *120, 121,* 160, *165*
Nanayakkara, N.P.D.   *119*
Nandy, A.K.   *121*
Navarro, E.   *119*
Neary, J.   *164*
Ngassapa, O.   *121*
Nicot, J.   *164*
Nilsson, K.   *120*
Nutakul, W.   *121*

O'Hagan, D.   *164*
Ortega, D.A.   *116*
Ortega, Z.   *116*
Ott, W.R.   *165*

Pant, P.   *121*
Paredes, A.   *121*
Patni, R.   *120*
Patra, A.   59, *121*
Patwardhan, S.A.   *116*
Patzelt, H.   *165*
Pavanand, K.   103, *121*
Pedersoli, J.L.   *122*
Pena, V.   *116*
Perales, A.   *118*
Peter, H.   126, *164*
Pezzuto, J.M.   59, *115, 121*

Pinheiro, J.A. *122*
Pomponi, M. *116, 122*
Prabhakaran, P.C. *165*
Premakumara, G.A.S. *122*
Probst, A. *164*

Ramanathan, J.D. *120*
Ramirez, G. *116*
Rangaswami, S. *120, 122*
Rastogi, R.P. *121*
Ratnasooriya, W.D. *122*
Ravelo, A.G. 19, *118, 119, 121–123*
Reddy, G.C.S. 104, *122*
Robinson, J.A. *165*
Robson, N. 19, *122*
Rodriguez, M.L. *118*
Rodriguez-Hahn, L. *116*
Roy, G. *121*

Salazar, J.A. *118*
Salisbury, P.J. *120*
Samsoedin, R. *164*
Sandeep, H. *165*
Sandmeier, P. 126, 127, 129, *164*
Satiro, M.W. *117*
Schechter, M.S. 4, 37, 79, *122*
Schwarting, A.E. *116*
Schwenk, E. 99, *122*
Seshadri, S. 39, 40, 77, 81, *122*
Seshadri, T.R. 71, *115, 120*
Shah, R.C. 4, 37–39, 43, 74, 82, *120, 122*
Shannon, J.S. *115*
Shieh, H.-L. *115*
Shirota, O. *120, 123*
Siegler, E.H. *120*
Silva, G.D.F. *122*
Simmonds, D.J. *115*
Sindelar, R.D. *120*
Singh, P. *120*
Singhi, C.L. *120*
Sneden, A.T. 59, 100, *122*
Soejarto, D.D. *121*
Sotheeswaran, S. *120*
Staunton, J. *165*
Staver, M.J. *165*
Steffan, B. *123*
Steglich, W. 113, *123*
Steiner, O. 156, *165*
Strobel, G.A. *164*

Strobel, S.A. *164*
Stuart, K.L. *120*
Suarez, E. *118*
Subramaniam, S. *117*
Sugawara, F. 127, 159, *164*
Sultanbawa, M.U.S. *119, 123*
Swanson, S.J. *165*
Swingle, M.C. *120*
Szendrei, K. *115*

Takahashi, N. *164*
Takahashi, Y. 71, *120, 121*
Takeya, K. *120, 123*
Tam, S.W. *120*
Tamm, Ch. 126, 127, 129, 156, *164*
Tan, W. *165*
Tezuka, Y. *116, 117, 119, 122*
Thakore, V.M. 82, *122*
Thomson, R.H. 39, 40, 71, 81, 93, 104, *115, 116, 120*
Tomioka, N. *120*
Townsley, P.M. *120*
Turner, A.B. *122, 123*

Vazquez, G. *118*
Viswanathan, N.I. *121, 123*

Wagner, H. 2, 19, 95, *115*
Walba, D.M. *120*
Walser-Volken, P. *165*
Watanabe, T. *123*
Wazeer, M.I.M. *119, 120*
Webster, H.K. *121*
Weeratunga, G. *116*
Weigert, E. *118*
Whiting, D.A. 74, *115, 120*
Wijeratne, D.B.T. *120, 123*
Wimalasiri, W.R. *119*
Woodland, D.W. *123*
Worth, B.R. *120*

Yamamoto, Y. *165*
Yang, C.C. *165*
Yongvanitchit, K. *121*
Yoshida, S. *164*
Yue, S. *165*

Zamudio, A. *116*

Subject Index

Acanthothamnus aphyllus 21–23, 25
Acetaldehyde 77
[1-^{13}C]Acetate 160
[1,2-^{13}C]Acetate 160
[1-^{14}C]Acetate 142
Acetic anhydride 76, 83
Acetone 36, 74, 82
19-O-Acetylchaetoglobosin A 136
Acetylcysteamine thioesters 142
Acetylene 158
S-Adenosyl-L-methionine 135, 150, 153, 154, 160
Agaricales 113
DL-[1-^{14}C]Alanine 135, 136
Alkylpicene 77
Aluminium 76
Antibacterial activity 4, 20, 35, 96
Antibiotic activity 35, 96, 97
Anticancer activity 96
Antihemorrhagic activity 103
Antileukaemic activity 24, 95
Antimalarial activity 103
Antimicrobial activity 96–98
Antiseptic activity 96
Antitumor activity 20, 95, 99–101
Arthropsadiol A 160–164
Arthropsadiol B 160, 161
Arthropsatriol A 160–162, 164
Arthropsatriol B 160, 161
Arthropsatriol C 160, 161
Arthropsatriol D 160, 161
Arthropsis truncata 160
Arthropsolide A 160–164
Arthropsolide B 160–162
Arthropsolide C 160, 161
Arthropsolide D 160, 161, 163, 164
L-[2-^{13}C,^{15}N]Aspartate 140
L-[U-^{14}C]Aspartate 135, 139
Aspartic acid 156, 161

L-Aspartic acid 139, 140, 156
L-[2,3-^{13}C$_2$]Aspartic acid 139
L-[U-^{14}C]Aspartic acid 135, 136
Atropocangorosin A 15, 20
Austroplenckia populnea 21, 22, 25

Bacillus alvei 97
Bacillus anthracis 97
Bacillus cereus 97, 98
Bacillus cereus mycoides 97
Bacillus megaterium 97
Bacillus mycoides 97
Bacillus pumilus 97
Bacillus subtilis 97, 98
Balaenol 10, 20, 36, 44, 46, 50, 56, 63, 64, 71
Balaenonol 7, 10, 20, 44, 46, 50, 56, 64, 71, 92, 93, 105
Benzamidine hydrochloride 150
Biological activity 2, 3, 15, 95
Borane 158
Boroacetic anhydride 76
Bromine 83
N-Bromosuccinimide 89
Brucella abortus 97
n-Butanol 36

(1S)-(−)-Camphanic chloride 158
Cangoronine 91
Cangorosin A 15, 20, 48
Cangorosin B 14–16, 20, 70, 101, 113
β-Carotene 3, 37
Cassine balae 20–23, 25, 92, 93
Cassine metabelica 22, 25
Cassine sp. 18
Catalytic hydrogen 79
Catha cassinoides 20–23, 25
Catha edulis 21–23, 25, 95

Celastraceae 2–4, 15, 18, 19, 24, 40, 93, 95,
97, 103, 105, 113
Celastrales 18
Celastranhydride 6, 18, 20, 70
Celastrol 3, 4, 6, 7, 8, 20, 24, 27, 35, 37–40,
42–44, 47, 51, 59, 61, 62, 70, 76–79, 81,
91, 93, 97, 99–103, 105
Celastroloids 2–7, 18–20, 24, 27, 35, 42,
43, 46–49, 56, 58, 59, 70–72, 74, 76, 79,
82, 83, 89–91, 93–95, 97–100, 102–106,
108, 109, 113
Celastrus articulatus 20
Celastrus dispermus 22, 25, 26, 36
Celastrus paniculatus 4, 20, 22, 24, 25, 35,
103
Celastrus scandens 3, 20, 25, 37
Celastrus sp. 18, 95
Celastrus strigillosus 20, 25
Cerin 40
Chloroform 36
Congorosin A 59
Congorosin B 59
Copper acetate 76
Corasil II 36
Crossopetalum uragoga 22, 23, 25
Cycloarthropsadiol C 160, 161
Cycloarthropsone 160, 161
Cytostatic activity 99, 100
Cytotoxic activity 100–102, 109, 113

Demethylzeylasteral 13, 21, 45
Demethylzeylasterone 13, 21, 45
Denhamia pittosporoides 22, 25, 36
21-Deoxydiacetyldihydrotingenone 29
21-Deoxydihydroexcelsinol 29
6-Deoxyerythronolide B 141
21-Deoxyexcelsine 28, 81, 82
Desmethylzeylasterone 21
Desoxycholic acid 83
Deuteromycetes 126
Diacetyldihydroiguesterin 30, 101
Diacetyldihydroisoiguesterin 30
Diacetyldihydroisotingenone-III 31
Diacetyldihydropristimerin 80
Diacetyldihydrotingenol 29
Diacetyldihydrotingenone 29, 101
Diacetylisopristimerin-III 81, 86
Diacetylisotingenone-III 33, 45
Diacetyl-6-oxopristimerol 89
Diacetylpristimerol 80, 81, 100, 101

Dibenzoyl peroxide 89
Dihydroatropocangorosin A 15, 21
Dihydrorussulaflavidin 113, 114
Dihydrotingenol 29
7′,8′-Dihydroxuxuarine Aβ 107, 109, 111,
112
3β,15α-Dihydroxyolean-12-ene 92
15α,22β-Dihydroxytingenone 7, 8, 21, 52,
60–62, 92
Dimeric celastroloids 15, 68
Dimeric phenolic triterpenoids 15
Dimeric quinonemethide-6-oxophenolic
celastroloids 79
Dimeric quinonemethide-6-oxophenolic
triterpenoids 16
Dimethyl acetonylpristimerol 82
Dimethyldihydroisoiguesterin-III 32
Dimethyldihydroisotingenol-III 31
Dimethyldihydronetzahualcoyone 100,
101
Dimethyl-21β-hydroxyisopristimerin-III
33, 45, 86, 87
Dimethylisocelastrol-III 33, 45
Dimethylisopristimerin-III 33
Dimethylisotingenone-III 33
Dimethyl-23-oxoisopristimerin-III 33
Dimethyl-6-oxopristimerol 34, 45
Dimethyl-23-nor-6-oxopristimerol 34
Dimethylpristimerol 29, 77
Dimethyl sulfate 82
Dimethyl sulfoxide 48
Dimethylzeylasteral 34, 45
Dimroth reagent 76
Diplococcus pneumoniae 96
Dispermoquinone 14, 21, 28, 46, 47, 58,
81, 82
Dithiothreitole 150
Drechslera tritici-repentis 159
Dulcitol 3, 19

EDTA 150
Ehrlich carcinomas 99
Elaeodendron balae 25
Elaeodendron glaucum 96
Elaeodendron sp. 19
9(11)-Enequinonemethides 5, 6, 43, 87,
104
14(15)-Enequinonemethides 5, 6, 43, 87,
91

9(11)-Enequinonemethide triterpenoids 11, 44
14(15)-Enequinonemethide triterpenoids 7, 10, 44, 56, 63, 64, 72
Erlenmeyer flask 127
Enterobacter hafniae 97
Epidermoid carcinoma 99
Erythromycin A 141
Escherichia coli 96–98, 139
Ethanol 88
Ethyl acetate 35, 36
Euonymus americanus 96
Euonymus purpureus 96
Euonymus sp. 18
Euonymus tingens 21, 23, 25, 35, 36
Excelsine 7, 8, 21, 51, 60–62, 101

Friedelanes 59, 67, 91
Friedelin 89, 91, 92
D:A-*Friedo-nor*-oleananes 2
D:A-*Friedo*-oleananes 2
[2,3-$^{13}C_2$]Fumaric acid 139
Fungi imperfecti 126, 160

Gibberella fujikuroi 97
D-Glucose 127
Glycollic aldehyde 77
Glyptopetalum sclerocarpum 21, 25
Gutta-percha 19
Gymnosporia emarginata 21–23, 25
Gymnosporia montana 21–23, 25
Gymnosporia sp. 19

Hippocratea aspera 25, 26
Hippocratea capulcensis 96
Hippocratea excelsa 21–23, 25
Hippocratea indica 26
Hippocratea sp. 19
Hippocrateaceae 2, 4, 18, 19, 24, 104, 105
Hydrogen chloride 83
30-Hydroxy-20(30)-dihydroisoiguesterin 103
Hydroxydroserone 37
16β-Hydroxyiguesterin 105, 106, 108
(20α)-3-Hydroxy-2-oxo-24-*nor*-friedela-1(10),3,5,7-tetraen-carboxylic acid-(29)-methylester 6
21β-Hydroxypristimerin 8, 21, 33, 44, 51, 85, 93

30-Hydroxypristimerin 7, 8, 21, 52, 61, 62, 94
Hydroxypristimerinene 11, 21, 44, 71
20-Hydroxytingenone 49
20α-Hydroxytingenone 7, 8, 21, 27, 35, 49, 52, 59, 61, 62, 93, 103, 104
22β-Hydroxytingenone 9, 11, 21, 27, 36, 44, 52, 59, 60, 63, 91–93, 100–103, 108, 109
20-Hydroxy-20-*epi*-tingenone 7, 8, 21, 24, 44, 49, 51, 61, 62, 93, 100–104
20α-Hydroxymaitenin 21
Hyphomycetes 126, 160

Iguesterin 7, 9, 21, 30, 44, 51, 93, 94, 99, 101, 105
Inhibitory activity 98, 103
Insecticidal activity 3, 4
Isobalaendiol 10, 21, 44, 50, 56, 64, 71, 103
Isobalaenol 10, 21, 36, 93
Isoiguesterin 7, 9, 21, 44, 54, 60, 63, 94, 100, 101
Isoiguesterol 9, 22, 54, 63, 103
Isopristimerin-I 47, 84, 85
Isopristimerin-II 84, 85
Isopristimerin-III 12, 22, 36, 45, 67, 84–89, 92, 101
Isopropyl ether 35
Isotingenone-III 12, 22, 36, 45, 67, 85, 88, 101

21-Ketopristimerin 7, 9, 22, 93
Klebsiella pneumoniae 96, 97
Kokoona ochracea 21, 23
Kokoona ochrasia 20, 25
Kokoona reflexa 20, 22, 24, 25
Kokoona zeylanica 14, 18, 20–22, 24, 25, 91, 96
Kokum soap 95, 96

Lamiaceae 19
Lipid-lowering activity 126
Lupanes 2
Lymphocytic leukemia P388 100
Lymphoepithelioma 99

Magellanin 107, 112, 113
Maitenin 9, 23, 29, 49, 81–83, 100
Malic acid 161, 162
D-Mannitol 127
Maytansine 24

Maytenin 9, 23, 97, 99, 100, 103
Maytenoquinone 46, 47
Maytenus boaria 22, 25, 96
Maytenus buchananii 24
Maytenus canariensis 25, 105, 106
Maytenus chuchuhuasca 22, 23, 25, 106,
    107–109
Maytenus dispermus 21, 22, 25, 26
Maytenus horrida 21–23, 26, 97
Maytenus ilicifolia 14, 20–23, 26, 36, 96,
    97, 106, 108
Maytenus krukovii 25
Maytenus laevis 21, 23, 26
Maytenus magellanica 107, 113
Maytenus nemerosa 21, 23, 26
Maytenus obtusifolia 21–23, 26
Maytenus rigida 21–23, 26
Maytenus senegalensis 25, 96
Maytenus sp. 15, 18, 19, 21, 23, 108
Maytenus umbellata 14, 20–24, 26
6-Mercaptopurine 100
Methanol 87
L-Methionine 135, 139, 140, 150
O-methyl-N-amino-propanal 158
Methylcelastrol 38
3-Methyl-22β,23-dihydroxy-6-oxo-
    tingenol 106, 108, 109
L-[Methyl-^{13}C]methionine 139
L-[Methyl-^{14}C]methionine 135, 136
Methymycin 141
3-Methyl-6-oxotingenol 106, 108, 109
(1R, 2S, 5R)-(−)-Methyl-(S)-4-
    toluenesulfinate 158
3-Methylxuxuarine Aα 111
Micrococcus luteus 97
Monensin A 141
Mortonia greggii 20, 26
Mortonia palmeri 20, 26
Mycobacterium phlei 97
Mycobacterium smegmatis 97

Narginin 141
Natural celastroloids 24, 25
Netzahualcoyene 10, 22, 44, 64, 71, 78, 87,
    88, 90, 97, 100, 101, 105
Netzahualcoyol 7, 10, 22, 44, 64
Netzahualcoyondiol 10, 22, 44, 64, 97,
    100, 101
Netzahualcoyone 7, 10, 22, 63, 75, 93, 97,
    98, 100, 101

Netzahualcoyonol 10, 22, 44, 97, 101
Nickel 76
23-*Nor*-6-oxodemethylpristimerol 13, 22,
    45
23-*Nor*-6-oxopristimerol 13, 22, 34

Orthosphenia mexicana 20, 22, 26
Orthosphenic acid 91
23-Oxoisopristimerin-III 12, 22, 33, 45,
    71, 73, 88, 91
6-Oxophenolic *nor*-triterpenoids 6
6-Oxophenolic triterpenoids 5, 6, 12, 14,
    43, 45, 47, 56, 57, 70
6-Oxopristimerol 106, 108, 109
7-Oxoquinonemethides 5, 6, 43
7-Oxoquinonemethide triterpenoids 14,
    47
6-Oxotingenol 106, 108, 109

Pachystima canbyi 22, 26
(3R)-Pentin-3-ol 158
Perchloric acid 84
Peritassa campestris 23, 26
Petroleum ether 35, 74
Phenolic triterpenoids 2, 5, 12, 14, 43, 45,
    67
L-Phenylalanine 136
Phenylmethanesulfonylfluoride 150
Phthiocol 37
Phytotoxic activity 159
Phytotoxins 159
Plasmodium falciparum 103
Plenckia populnea 22, 23, 26
Plumbagin 37, 99
Polpunonic acid 91, 97
Polyisoprene 19
Primin 99
Prionostemma aspera 21, 23, 25, 26
Pristimeria grahamii 23, 26
Pristimeria indica 4, 23, 26, 35, 37, 96
Pristimerin 3, 4, 6, 7, 9, 22, 24, 28, 29,
    33–44, 46–50, 54, 58–60, 63, 70, 71,
    74–88, 90–94, 96, 97, 99–101, 103, 105,
    108, 109, 113, 114
Pristimerinene 11, 23, 71, 87, 88, 104, 105
Pristimerin leucotriacetate 81, 83, 86, 89
Pristimerol 40, 80, 99
Pristimerol bis-*p*-bromobenzoate 74, 75
Pristimerol dimethyl ether 40
1,3-Propanedithiol 158

Propionaldehyde 158
Proteus mirabilis 97
Proteus vulgaris 96
Pseudomonas aeruginosa 97
Pseurotin A 136
Pummerer reaction 158
Pyridine 83
Pyridinium dichromate 158

Quinonemethides 5, 43, 52, 54, 61, 63
Quinonemethide triterpenoids 6, 7, 8, 11, 14, 44, 51, 71

Raney nickel 79
Reissantia grahamii 26
Reissantia indica 20, 23, 26
Reissantia sp. 19
Russula flavida 113
Russulaflavidin 113, 114
Rzedowskia bistriterpenoid-I 17, 23, 58, 68, 70, 78, 90, 98
Rzedowskia bistriterpenoid-II 17, 23, 58, 68, 70, 78, 90, 98
Rzedowskia tolantonguensis 21–23, 26

Saccharomyces cerevisiae 97, 98
Salacenonal 91
Salacia campestris 26
Salacia crassifolia 23, 26
Salacia macrosperma 19, 21, 23, 24, 26
Salacia madagascariensis 21, 26
Salacia reticulata var. diandra 21, 26
Salacia reticulata var. β-diandra 20–23, 26, 91, 94
Salacia sp. 18, 19, 21–23, 26, 27
Salacia quinonemethide 11, 23, 44, 104, 105
Salaciquinone 7, 14, 23, 58, 70
Salaspermic acid 91, 97
Salmonella paratyphii 96
Salmonella sp. 97, 98
Salmonella typhii 96
Saptarangi quinone-A 19
Sarcina lutea 97
Schaefferia cuneifolia 23, 24, 27, 93, 97
Schizontocidal activity 103
Sephadex LH-20 36, 105, 113
DL-[1-^{14}C]Serine 135, 136
Shigella dysenteriae 96

Silica gel 35, 36, 105, 113, 129, 143, 147
Siphonodon australe 40
Skelly-solve B 35
Sodium [1-^{13}C]acetate 136
Sodium [1, 2-^{13}C$_2$]acetate 138
Sodium [1-^{14}C]acetate 135
Sodium [2-^{14}C]acetate 135
Sodium bisulfite 76, 83
Sodium borohydride 76
Sodium deoxycholate 83
Spermicidal activity 103
Spirostaphylotrichin A 126–131, 134–142, 144, 149–157, 159
Spirostaphylotrichin B 129–131, 134, 144, 146, 149–152, 159
Spirostaphylotrichin C 127, 129–131, 134, 149–155, 159, 161
Spirostaphylotrichin D 127, 129–131, 149–155, 159, 161
Spirostaphylotrichin E 132, 134, 143–145, 153
Spirostaphylotrichin F 129, 131, 132, 134, 143–145, 153
Spirostaphylotrichin G 132, 143–145, 149–157
Spirostaphylotrichin H 132, 143–145, 149–157
Spirostaphylotrichin I 132, 143–145, 153
Spirostaphylotrichin K 133, 134, 143–146, 153, 161
Spirostaphylotrichin L 133, 134, 143–145, 153
Spirostaphylotrichin M 133, 134, 143–145, 153
Spirostaphylotrichin N 146–152
Spirostaphylotrichin O 142, 146–150, 153, 154
Spirostaphylotrichin P 146–150, 153–156
Spirostaphylotrichin Q 129–131, 134, 153
Spirostaphylotrichin R 129–131, 134, 153, 159
Spirostaphylotrichin S 133, 134, 143–145, 153
Spirostaphylotrichin T 146–148, 153
Spirostaphylotrichin V 146–154
Staphylococcus aureus 96–98
Staphylococcus epidermidis 97
Staphylococcus saprophyticus 97
Staphylococcus warneri 97

Staphylotrichum coccosporum 126–129, 131, 135, 136, 138, 142, 150, 151, 153
Stomach carcinoma 99
Streptococcus faecalis 96, 97
Streptococcus haemolyticus 97
Streptococcus pyrogenes 96, 97
Streptococcus viridens 96
Streptomycetes 142
$[2,3\text{-}^{13}C_2]$Succinate 139
Sulfur dioxide 80
Sulfuric acid 76, 83, 84
Sulfuric ether 35
Sulfurous acid 76, 79

Tetramethylchrysene 38
Tetronasin 141
Thiele reagent 76
Tingenin A 9, 23, 36, 59
Tingenin B 9, 21, 36, 59
Tingenone 7, 9, 23, 24, 27, 29–33, 35, 36, 44, 47, 49, 54, 59, 63, 71, 74, 78, 80–83, 85, 86, 90, 93, 97, 99–103, 108, 109
Toluene-*p*-thiol 82
Triacetyldihydroisotingenol-III 31
Triacetyldihydromaitenol 29
Triacetyldihydrotingenol 29
Trichloroacetyl isocyanate 110
Triethylamine 80
Trimethylzeylasterone 34, 45, 90
1,2,5-Trimethyl-6,7-dihydroxynaphthalene 86
1,2,6-Trimethylphenanthrene 77
Tripterine 3, 4, 8
Tripterygium regelii 20, 27
Tripterygium wilfordii 3, 4, 20, 27
Triptolide 27
Triterpene phenoldienones 2

Triticone A 159
Triticone B 159
Triticone C 159
Triticone D 159
Triticone E 159, 160
Triticone F 159, 160
DL-Tryptophan 136
Tylactone 141

Umbellatin α 14, 17, 24, 68, 70, 78, 90, 98, 109, 110
Umbellatin β 17, 24, 78, 90, 98, 109, 110

Vibrio cholerae 96
Vitamin A 3
Vitamin K_1 37
Vitamin K_2 37

Xuxuarine Aα 106, 109–112
Xuxuarine Aβ 106, 109, 112
Xuxuarine Bα 106, 109, 110
Xuxuarine Bβ 107, 109, 112
Xuxuarine Cα 107, 109–111
Xuxuarine Cβ 107, 109, 111, 112
Xuxuarine Dα 107, 109, 110
Xuxuarine Dβ 107, 109, 112
Xuxuarines 110–113

Yoshida sarcomas 99

Zeylanol 91
Zeylasteral 13, 24, 34, 45, 56, 57, 67, 70, 91, 92
Zeylasterone 13, 14, 24, 34, 45, 46, 56, 57, 66, 67, 70, 73, 90–92, 103
Zinowiewia integerrima 23, 27

SpringerChemistry

Fortschritte der Chemie organischer Naturstoffe

Progress in the Chemistry

of Organic Natural Products

Founded by L. Zechmeister
Edited by W. Herz, G. W. Kirby, R. E. Moore, W. Steglich,
and Ch. Tamm

Volume 66

1995. 6 figures. VII, 332 pages. Cloth DM 290,–, öS 2030,–
Subscription price: Cloth DM 261,–, öS 1827,–. ISBN 3-211-82597-5

Contents:
T. Okuda, T. Yoshida, T. Hatano: Hydrolyzable Tannins and
Related Polyphenols.
R. G. de Souza Berlinck: Some Aspects of Guanidine Secondary
Metabolites.

Volume 65

1995. 2 figures. IX, 618 pages. Cloth DM 440,–, öS 3080,–
Subscription price: Cloth DM 396,–, öS 2772,–. ISBN 3-211-82576-2

Contents:
Y. Asakawa: Chemical Constituents of the Bryophytes.

 SpringerWienNewYork

P.O.Box 89, A-1201 Wien • New York, NY 10010, 175 Fifth Avenue
Heidelberger Platz 3, D-14197 Berlin • Tokyo 113, 3-13, Hongo 3-chome, Bunkyo-ku

SpringerChemistry

Fortschritte der Chemie organischer Naturstoffe
Progress in the Chemistry
of Organic Natural Products

Founded by L. Zechmeister
Edited by W. Herz, G. W. Kirby, R. E. Moore, W. Steglich, and Ch. Tamm

Volume 64

1995. 22 partly coloured figures. VII, 216 pages. Cloth DM 250,–, öS 1750,–
Subscription price: Cloth DM 225,–, öS 1575,–. ISBN 3-211-82533-9

Contents:
A. G. González and J. Bermejo Barrera: Chemistry and Sources of Mono- and Bicyclic Sesquiterpenes from Ferula Species.
G. Prota: The Chemistry of Melanins and Melanogenesis.
H. J. M. Gijsen, J. B. P. A. Wijnberg, and Ae. de Groot: Structure, Occurrence, Biosynthesis, Biological Activity, Synthesis, and Chemistry of Aromadendrane Sesquiterpenoids.

Volume 63

1994. VII, 216 pages. Cloth DM 220,–, öS 1540,–
Subscription price: Cloth DM 198,–, öS 1386,–. ISBN 3-211-82443-X

Contents:
A. B. Ray and M. Gupta: Withasteroids, a Growing Group of Naturally Occurring Steroidal Lactones.
L. Rodríguez-Hahn, B. Esquivel, and J. Cárdenas: Clerodane Diterpenes in Labiatae.

 SpringerWienNewYork

P.O.Box 89, A-1201 Wien • New York, NY 10010, 175 Fifth Avenue
Heidelberger Platz 3, D-14197 Berlin • Tokyo 113, 3-13, Hongo 3-chome, Bunkyo-ku

Springer-Verlag
and the Environment

WE AT SPRINGER-VERLAG FIRMLY BELIEVE THAT AN
international science publisher has a special obliga-
tion to the environment, and our corporate policies
consistently reflect this conviction.

WE ALSO EXPECT OUR BUSINESS PARTNERS – PRINTERS,
paper mills, packaging manufacturers, etc. – to commit
themselves to using environmentally friendly materials and production processes.

THE PAPER IN THIS BOOK IS MADE FROM NO-CHLORINE
pulp and is acid free, in conformance with international standards for paper permanency.